French Air Force

Contemporary Aircraft

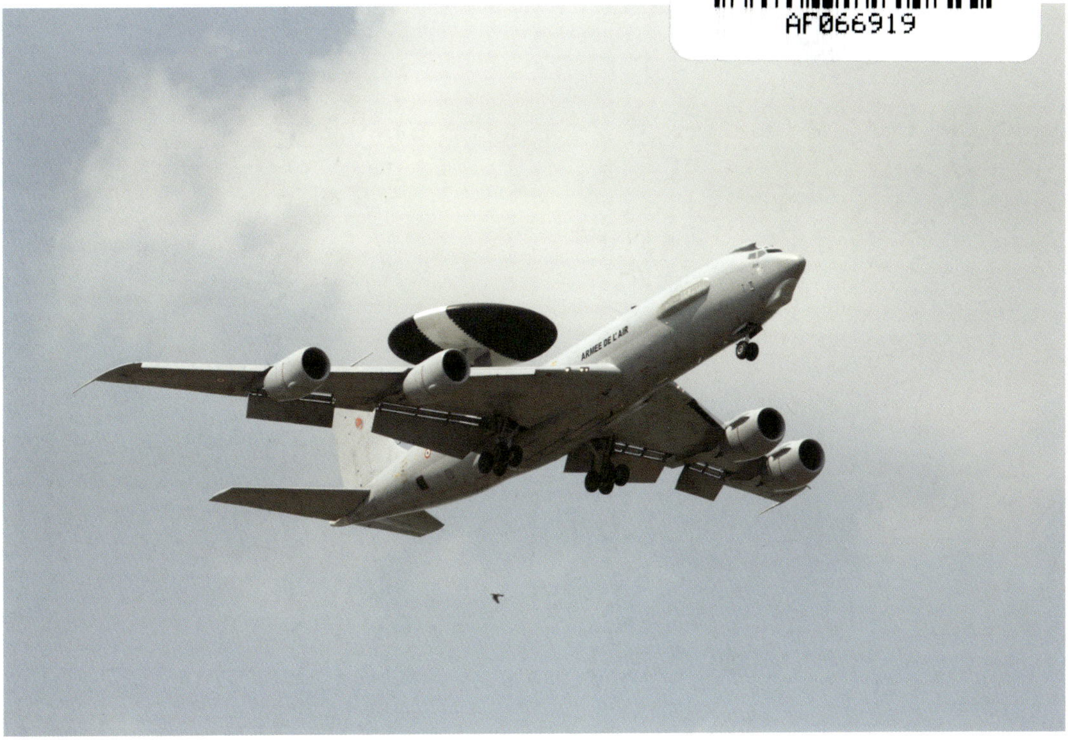

HENRI-PIERRE GROLLEAU

Key Books

AIR FORCES SERIES, VOLUME 7

Front cover image: A pair of Rafales aggressively pulls into the vertical. The aircraft in the foreground is fitted with a Scalp stealth cruise missile under the centreline pylon.

Title page image: Undercarriage and flaps fully down, this E-3F Sentry undertakes a slow pass during an airshow.

Contents page image: Rafale C single-seaters regularly take part in the French homeland air-defence mission. This QRA Rafale is armed with four live Mica RF/IR missiles.

Back cover image: The A400M has a cargo hold large enough to accommodate the latest generation of French Army armoured vehicles that are heavier, wider and taller than their predecessors.

Acknowledgements

The author would like to thank all military and civilian personnel for the invaluable assistance provided during his various reports and flights with the Armée de l'Air / Armée de l'Air et de l'Espace over the years.

Published by Key Books
An imprint of Key Publishing Ltd
PO Box 100
Stamford
Lincs PE9 1XQ

www.keypublishing.com

The right of Henri-Pierre Grolleau to be identified as the author of this book has been asserted in accordance with the Copyright, Designs and Patents Act 1988 Sections 77 and 78.

Text copyright © Henri-Pierre Grolleau, 2023
Photographs copyright © Henri-Pierre Grolleau, 2023, unless otherwise stated

ISBN 978 1 80282 426 1

All rights reserved. Reproduction in whole or in part in any form whatsoever or by any means is strictly prohibited without the prior permission of the Publisher.

Typeset by SJmagic DESIGN SERVICES, India.

Contents

Introduction	A Constantly Evolving Force	4
Chapter 1	Aircraft in the Inventory	16
Chapter 2	Command and Control / Intelligence-Collecting Aircraft	42
Chapter 3	Tankers and Airlifters	52
Chapter 4	Helicopters	80
Chapter 5	Trainers	98
Chapter 6	Surface-to-Air Missile Systems	120
Glossary		128

Introduction
A Constantly Evolving Force

The Armée de l'Air et de l'Espace (AAE, the French Air and Space Force) is constantly investing to renew its inventory of fixed-wing and rotary type aircraft. New, more modern aircraft are regularly introduced to bolster operational capabilities while others are retired after a long and successful career. Since 2000, a large number of types have been withdrawn from Armée de l'Air use including:

Mirage F1C single-seat interceptor, in 2003
DC-8 Sarigue electronic intelligence aircraft, in 2004
Nord 262 light transport aircraft, in 2004
Alouette III helicopter, in 2004
Jaguar A/E strike fighter, in 2005
Mirage IVP strategic reconnaissance aircraft, in 2005
Mystère 20 calibration, training and light transport aircraft, in 2007
EMB-312F Tucano basic trainer, in 2009
A319CJ corporate jet, in 2010
Mirage F1CT strike fighter, in 2012
Falcon 50 trijet business jet, in 2014
Mirage F1CR/B fighter reconnaissance aircraft / operational trainer, in 2014
Super Puma SAR helicopter, in 2016
Harfang unmanned aerial vehicle, in 2018
Mirage 2000N nuclear strike fighter, in 2018
TB30 Epsilon basic trainer, in 2019
Airbus A340 airliner, in 2021
Airbus A310 airliner, in 2022
Transall / Transall Gabriel tactical airlifter and its electronic intelligence variant, both in 2022
Mirage 2000C single-seat interceptor, in 2022.

It is fair to say that, since the end of the Cold War, in 1989, the French Air Force inventory has been shaped by budget constraints and by the successive rounds of cuts imposed by government. The switch from conscription to an all-professional force has had an impact too, helping to create a more powerful and much better trained force, but also a more costly organisation to run, had its original size been maintained. Obviously, the move away from low-wage conscripts to much higher-paid professionals has had a budgetary impact on procurement and on the replacement of older aircraft types, and was a major driver for down-sizing.

Multi-role platforms
The main focus behind the French Air Force's modernisation was the procurement of more capable platforms that could each carry out a larger range of roles than their predecessors. A typical example

Rafale B and C fighters refuelling from an A330 Phénix. The back seat of the Rafale B two-seater is empty.

is the Rafale omnirole fighter that has progressively become the main asset of the AAE's fighting force. It was developed so that just one aircraft could perform a wide array of missions that were previously allocated to a whole range of fighters. When designed, it was intended to replace the French Navy's F-8P Crusader carrier-borne interceptors, Étendard IVPM carrier-borne recce birds, Super Étendard strike fighters, the French Air Force's Jaguar A/E strike fighters, Mirage F1C/CT air-defence fighters, Mirage F1CR tactical reconnaissance fighters, Mirage IVP strategic reconnaissance aircraft and, eventually, Mirage 2000C/B interceptors and Mirage 2000N nuclear strike fighters. As a result, the new type was conceived to excel in an extremely wide range of roles from land bases and from the deck of aircraft carriers: nuclear deterrence, air-defence, offensive counter air, sweep, escort, destruction of enemy air defences, battlefield air interdiction, deep precision strikes, close air support, antiship attacks, pre-strategic and tactical reconnaissance, buddy-buddy in-flight refuelling…

The same concept applies to the A400M Atlas airlifter and to the A330 Phénix: the A400M can perform both tactical and strategic airlift missions, including assault landings on unprepared strips, while the Phénix is designed to excel in the in-flight refuelling, long-range transport of personnel, and cargo and medical evacuation roles. The A400M has now totally replaced the much-loved Transall. Until very recently planned for 2023, the retirement of the Transall was brought forward to 2022 to reduce costs. The twin had flown well past its retirement date and had provided sterling service to the French Armed Forces.

The future

The AAE is actively preparing for the future and new aircraft are already on the horizon. The retirement of the Transall Gabriel Electronic Intelligence (ELINT) platform in 2022 has left a capability gap that is being addressed by acceptance into service of the three Ceres (CapacitÉ de Renseignement Electronique Spatial, or space-based electronic intelligence capability) ELINT satellites launched in November 2021 and by the introduction, in 2025, of the Dassault Aviation Falcon 8X Archange long-range ELINT aircraft. The three Falcon 8X Archanges will be fitted with the CUGE (Capacité Universelle de Guerre Électronique, or universal electronic warfare capability) suite provided by Thales. They will be operated from Base Aérienne 105 Évreux-Fauville by Escadron Électronique Aéroporté 1/54 'Dunkerque' and will provide strategic-level electronic-gathering capabilities. Compared to the Transall Gabriel, they will be faster, will offer a much longer range and will fly much higher, thus pushing further into hostile territories the emitter detection reach, or allowing the aircraft to operate at stand-off distances from a threat.

The new Airbus Helicopters H160M Guépard, a derivative of the civilian H160 developed for the offshore VIP and SAR markets, will replace the Pumas and Fennecs and supplement a growing inventory of EC725/H225M Caracals. The Guépard will be extremely well adapted to the AAE's missions. For example, it will be equipped with an in-flight refuelling probe to extend both reach and mission duration, allowing it to escort Caracals engaged on long-distance CSAR or special operations missions.

France is also part of the Eurodrone programme, a highly controversial programme that will produce a massive twin-engine unmanned aerial vehicle (UAV) that will offer excellent surveillance capabilities, but at high procurement and operating costs. Moreover, this sort of slow and large UAV would prove extremely vulnerable in high-intensity combat scenarios.

Looking further ahead into the 2040s, the New Generation Fighter (NGF) will replace, at some stage, the Rafale. This future design will have to be stealthy, with an internal weapons' carrying capability. It will be bigger than the Rafale and is expected to be in the 30-tonne league; in comparison the Rafale weighs 24.5 tonnes at maximum weight. Like its predecessor, the NGF will have to be a fully multirole fighter capable of operating from aircraft carriers.

Satellites

As might be expected, the French Air and Space Force operates a whole array of space surveillance and space-based surveillance systems. France has long invested in satellites and space launchers. The European space port is located in Kourou, French Guiana, from where successive variants of the Ariane rockets have been launched for decades. Three types of satellites are in current use by the AAE:

Communication satellites of the Syracuse series that comprises Syracuse I (three Telecom 1A / 1B / 1C satellites launched between 1984 and 1987), Syracuse II (with four Telecom 2A / 2B / 2C / 2D satellites launched between 1991 and 1996), Syracuse III (two Syracuse 3A / 3B in 2005/2006 and the Franco-Italian Sicral 2 in 2015) and Syracuse IV (the first of which, Syracuse 4A, was sent into orbit in 2021 to be followed by 4B in 2023). Not to be forgotten is the Italo-French Athena-Fidus, launched in 2014.

Reconnaissance satellites of the Helios family (with Helios 1A and 1B launched in 1995 and 1999 and Helios 2A and 2B in 2004 and 2009), Pléiades (1A in 2011 and 1B in 2012) and Composante Spatiale Optique (optical space-based component, with CSO 1 in 2018, CSO 2 in 2020 and CSO 3 to be launched in 2023).

Electronic intelligence-gathering satellites developed over a long period, with three generations of demonstrators (Cerise, Essaim 1 to 4, and Elisa 1 to 4, respectively orbited in 1995, 2004, and 2011) followed by the Ceres ELINT constellation composed of three satellites sent into orbit in 2021 to triangulate the position of radar emitters.

The French Air and Space Force fields two fighter families, the ageing Mirage 2000 (seen here leading) and the state-of-the-art Rafale.

The EC725 Caracal has proved well adapted to the needs of the French Armed Forces. This Caracal is not fitted with an in-flight refuelling probe.

The French C-130H/H-30 fleet plays a crucial role within the French air mobility assets, helping move troops, equipment and supplies in Europe, Africa and further afield.

In full recognition of rising threats, the French Air and Space Force is currently strengthening its force protection units.

French Air and Space Force aircrews are all trained to very high standards. They have amassed a wealth of experience during countless combat missions in the Gulf War, and in Bosnia, Kosovo, Afghanistan, Libya, Sahel and Iraq.

This Escadron d'Hélicoptères 5/67 'Alpilles' Fennec has intercepted a Cessna and forced it to land at Orange during an exercise. The squadron was recently renumbered 1/65 'Alpilles'.

This photo clearly shows the difference in size between the C-135FR Stratotanker (leading) and the new A330 MRTT Phénix tanker.

One C-135FR Stratotanker is withdrawn from use for each new A330 MRTT Phénix that enters service.

The Alpha Jets of the Patrouille de France are a common sight all over Europe. New manoeuvres are regularly choreographed by the team to improve the demonstration and increase its appeal to the public.

A modern air force requires mobile command structures and the Armée de l'Air et de l'Espace is no exception. Here, a mobile command post set up in the field.

French Air and Space Force aircrews are trained to operate day and night, in all sorts of environments and weather conditions.

The Casa CN235 provides sterling service in continental France, in French overseas territories and in Africa. The Casa CN235 has become the AAE's workhorse for all light transport missions in continental France and abroad.

This photo shows the difference in size between the now withdrawn from use Transall (foreground) and the CN235.

The French Air and Space Force boasts a dedicated para team that displays at airshows in France and further afield.

Super Hercules (foreground) and Atlas, the latest two airlifters to have entered service in France. This shot shows how big the A400M Atlas really is compared to the C-130J/KC-130J family.

An A400M Atlas and a C-130J-30 share the ramp at Base Aérienne 105 Évreux, in May 2022.

This 'Normandie-Niémen' Rafale is configured for deep strike missions with a Scalp stealth cruise missile under its belly.

Chapter 1

Aircraft in the Inventory
Fighters

Dassault Aviation Mirage 2000B

The Mirage 2000 programme was launched in the 1970s to produce a replacement for the Mirage III and Mirage F1 fighters which, alongside the Mirage 5 and the Jaguar strike fighter, formed the backbone of the French Air Force. The end result was a sleek lightweight delta fighter powered by a Snecma M53-5 turbofan (rated at 9,000kg / 19,842lb of thrust in full reheat) and equipped with the RDM (Radar Doppler Multifonction) multimode Doppler radar. The new aircraft was the first French fighter to reach full-scale production to be fitted with fly-by-wire (FBW) controls for maximum agility. With its highly swept wings, it was optimized for high speed but, thanks to its advanced FBW controls, it could also remain extremely agile and fully controllable at incredibly low speeds, as routinely demonstrated at airshows. The delta fighter's swept wings and low-drag airframe coupled with its high thrust were decisive in providing short take-off distances, excellent acceleration and outstanding climb rates.

The Mirage 2000 was eventually developed into four main variants for the French Armed Forces: the Mirage 2000C single-seat interceptor, the Mirage 2000B two-seat trainer, the Mirage 2000N nuclear-deterrence strike fighter and the Mirage 2000D conventional strike fighter. An improved variant, the Mirage 2000-5F interceptor, was an upgrade produced from earlier Mirage 2000C aircraft. In French Air Force parlance, the air-defence Mirage 2000C/B/2000-5F pilots are known as the 'Bleus' (blues), in reference to their paint scheme, while their Mirage 2000D and N colleagues are the 'muds', for 'mud movers', a common nickname for ground-attack aircraft within NATO.

The first 37 Mirage 2000C and B aircraft to come out of the assembly line were equipped with the RDM radar, but production soon switched to the improved RDI (Radar Doppler d'Interception) that offered longer detection and tracking performance against airborne targets. The RDI then provided French pilots with impressive look-down shoot-down capabilities when used in conjunction with the long-range, radar-guided, semi-active Super 530D missile. The early M53-5 turbofan that equipped

An anonymous Mirage 2000B comes in to land at Orange at the end of a training sortie, in 2007.

the first 37 Mirage 2000s also gave way from production aircraft No 38 to the more powerful M53-P2 turbofan variant rated at 9,700kg (21,385lb) of thrust at full power.

A total of 30 Mirage 2000B trainers were delivered to the French Air Force, but only seven remain in service at the time of writing following a number of crashes, the early retirement of M53-5-powered aircraft and the sale of a further two to Brazil. The two-seat variant is mainly used for the first phases of the conversion training programme, allowing instructors to fly with young pilots straight out of advanced training, with seasoned aircrews converting from another fast jet type, or with experienced Mirage 2000 pilots coming back to a front-line squadron after a ground tour. The two-seat Mirage 2000B also plays a crucial role providing in-flight refuelling training for all Mirage 2000D/-5F pilots. Like the Mirage 2000C, the Mirage 2000B could undertake combat missions with Super 530D semi-active radar-guided long-range missiles and with Magic 2 fire-and-forget infrared-guided short-range air-to-air missiles. However, while the Mirage 2000B could carry externally the same weapons as the single-seater, it was not fitted with the two DEFA 554 30-mm internal cannons that equipped the Mirage 2000C. The guns and their ammunition boxes had been removed to make room for black boxes that had to be moved to give way to the 2000B's rear cockpit and its additional Martin-Baker Mk 10 ejection seat. Although it was very rarely seen, the Mirage 2000B could still carry the CC420 30-mm gun pod under its centreline pylon. While the Mirage 2000B was theoretically capable of carrying air-to-ground ordnance, this was an extremely rare occurrence. The withdrawal from service of the Super 530D missile in 2010 only left Magic 2 missiles available for the Mirage 2000C and Mirage 2000B, severely curtailing their air-to-air capabilities.

After the official disbandment of EC2/5 'Ile-de-France', in June 2022, the last of the 124 Mirage 2000C single-seaters produced for the Armée de l'Air had all been withdrawn from use by the end of July 2022, leaving only the Mirage 2000B and the Mirage 2000-5F as the last members of the Mirage 2000 air-defence community in service in France. 'Ile-de-France' is expected to be recreated at Orange as a Rafale squadron in 2024/2025. The seven remaining Mirage 2000B two-seaters moved from Base Aérienne 115 Orange-Caritat to Base Aérienne 133 Nancy-Ochey where they will continue to be used in the conversion training role by Escadron de Chasse 2/3 'Champagne' (a Mirage 2000D squadron) for the foreseeable future. A large percentage of EC 2/5 instructors have been posted away to EC 2/3 in an effort to retain the expertise within the Mirage 2000 community. The Mirage 2000C/B RDI simulator, located in Orange for years, has also been moved to Nancy after construction work has been carried out there to accommodate it. The two-seat Mirage 2000B will still be a stepping stone for all future Mirage 2000-5F and Mirage 2000D pilots for initial conversion training and for in-flight refuelling training.

While stationed in Orange, Mirage 2000Bs routinely operated from hardened aircraft shelters.

All French Mirage 2000 pilots have undergone early conversion work on the Mirage 2000B two-seater.

In full afterburner, Mirage 2000B no 520 takes off from Orange, in 2007. The two-seat Mirage 2000B is fitted with the same RDI radar as the single-seat 2000C, which was withdrawn from use in 2022.

Under the close supervision of an instructor, a young pilot has successfully plugged his probe into the basket of a C-135FR Stratotanker.

The Mirage 2000B was long used to support operational evaluation work from Mont-de-Marsan. Escadron de Chasse et d'Expérimentation 'Côte d'Argent' relinquished its last 2000B in the early 2010s.

Dassault Aviation Mirage 2000-5F

With its Mica fire-and-forget missiles, its RDY multimode Doppler radar and its reworked cockpit, the Mirage 2000-5F specialises in air-defence / air-superiority missions. All the Mirage 2000-5F air-defence fighters started their lives as Mirage 2000C interceptors. When Dassault Aviation launched the Mirage 2000-5 programme to conquer the export market, the French Ministry of Defence decided to support the effort by upgrading 37 of the 124 Mirage 2000C fighters delivered to the French Air Force to the much more capable 2000-5F standard. This update was also needed to provide increased air-defence / air-superiority capabilities at a time when the Rafale programme had been delayed after the end of the Cold War. It was felt that the Super 530D semi-active radar-guided missiles that equipped the Mirage 2000C would soon become outdated, overshadowed by more modern radar-guided fire-and-forget missiles such as the American AIM-120 AMRAAM (Advanced Medium Range Air-to-Air Missile). This was the main reason behind the development of the RDY radar and of the Mica missile that would transform the Mirage 2000 into an even more lethal and more capable interceptor.

The scope of the Mirage 2000-5F update was extremely large and the programme was complex, with an entirely new weapon system centred around the RDY radar specifically designed for air-defence missions. The RDY radar was developed by Thales as a successor to the RDI, but with considerably improved detection and tracking capabilities. While the RDI was optimised for the employment of the semi-active Super 530D missile, the RDY was intended to work in conjunction with the Mica family of air-to-air, fire-and-forget missiles. For air-to-air missions, its detection ranges and multi-target capabilities enable the pilot to track up to eight targets in the track-while-scan mode (out of 24 displayed), irrespective of their aspects and flying altitudes. Interception data are calculated for the four priority targets, allowing the firing of four Mica missiles in quick succession in their full range envelopes.

The Mica (Missile d'Interception, de Combat et d'Autodéfense, or interception, combat and self-defence missile) was conceived as a multi-role fire-and-forget missile that could be fitted with either an active radar seeker (Mica RF, for radiofrequency) or a passive infrared seeker (Mica IR). The Mirage 2000-5F was designed to ripple fire up to four Mica missiles at four different targets before turning away from the threat, a major advantage compared to the earlier Mirage 2000C/B that relied on semi-active Super 530D missiles that had to be guided until impact.

To be used in conjunction with the new weapon system, an entirely new cockpit was designed with five displays (hence the name 2000-5 / 'Dash 5'). In fact, three of the displays are traditional screens while the fourth is collimated to infinity and the fifth one is the Head-Up Display (HUD). The Mirage 2000-5F retains the same electronic warfare/self-protection suite as the earliest Mirage 2000C.

The quickest identifying feature on the Mirage 2000-5 compared with earlier variants is the lack of the traditional pitot tube that has been replaced by a side probe air-data system, such as on the Mirage 2000D. This choice was dictated by a need to increase radar performance because the nose-mounted pitot tube was an obstacle to proper radar wave propagation in the forward sector. All Mirage 2000C airframes selected for Mirage 2000-5F conversion were fitted with the M53-P2 turbofan variant rated at 9,700kg (21,385lbs) of thrust at full afterburner power.

The Mirage 2000-5F was initially armed with the Mica RF and with the Magic 2 short-range infrared-guided missile, but the latter was eventually supplanted by the longer-range and more modern Mica IR. The 2000-5F was also upgraded with the Link 16 datalink to increase its ability to share and receive tactical information and plug into French and NATO command-and-control networks.

Initially, three squadrons operated the 'Dash 5': Escadron de Chasse et d'Expérimentation 5/330 'Côte d'Argent'; the operational evaluation unit at Base Aérienne 118 Mont-de-Marsan; and

Escadrons de Chasse 1/2 'Cigognes' et 2/2 'Côte d'Or', both at Dijon-Longvic. Following, the closure of Dijon and the withdrawal of the Mirage F1C, at Djibouti, two squadrons are currently equipped with the Mirage 2000-5F: Escadron de Chasse 1/2 'Cigognes', at Base Aérienne 116 Luxeuil Saint-Sauveur, in the east of France, and Escadron de Chasse 3/11 'Corse', at Base Aérienne 188 Djibouti, in the Republic of Djibouti.

Right: Like all versions of the Mirage 2000, the Mirage 2000-5F is said to be remarkably agile, as demonstrated at numerous airshows over the years.

Below: An EC 1/2 'Cigognes' Mirage 2000-5F comes in to land at Cold Lake at the end of a training sortie during Maple Flag 2016.

A quartet of Mirage 2000-5Fs and four Rafales share the ramp at Cold Lake during Maple Flag 2016 exercise.

Three Mirage 2000-5F interceptors in close formation during Maple Flag 2016.

Close-up on a Mica RF fire-and-forget radar-guided air-to-air missile under a Mirage 2000-5F's forward left fuselage hardpoint.

Armed with live Mica missiles, this QRA Mirage 2000-5F has just intercepted an Alpha Jet for training purposes.

Armed with two live Mica RF/IR missiles, this Mirage 2000-5F breaks away from the Phénix tanker during an air-defence mission over Poland, in April 2022.

Dassault Aviation Mirage 2000D

The Mirage 2000D was developed as a conventional strike fighter from the earlier Mirage 2000N nuclear strike variant. Like the 2000N, the 2000D was optimised from the outset for very high-speed, very low-level penetration using an advanced terrain-following system composed of an Antilope terrain-following radar and two extremely precise inertial navigation systems. Like its predecessor, it was fitted with a fully internal electronic warfare / self-defence suite that included the Caméléon radar jammer and flare and chaff dispensers. A total of 86 Mirage 2000Ds were delivered to the Armée de l'Air from 1993 to 2001. All Mirage 2000D strike fighters are fitted with the M53-P2 turbofan variant rated at 9,700kg (21,385lbs) of thrust at full power.

In comparison to the earlier Mirage 2000N, the 2000D can be easily identified at first glance thanks to its green radome without a pitot tube (it was black, with a pitot tube on the N) and to the upward-firing decoy dispensers located on the spine of the aircraft, between the rear cockpit and the base of the fin. The 2000D also has an integral dark grey / dark green wraparound camouflage. while the underside of the 2000N was light grey.

Since becoming operational, the Mirage 2000D has been the specialised asset for all-weather conventional attacks with guided or unguided weapons, proving its worth operationally in Kosovo, Afghanistan, Libya, the Middle East and the Sahara. The fighter was initially equipped with French-built AS30L laser-guided air-to-surface missiles and BGL-1000 laser-guided bombs, and with an assortment or free-fall and retarded sleek bombs. Nowadays, US-sourced GBU-12, GBU-16, GBU-24 and GBU-49 of the Paveway II, Paveway III and Enhanced Paveway families are mainly used. Although unguided weapons were considered outdated after Gulf War 1, in 1991, they have been employed operationally in the Sahara for saturation attacks over large areas, with Mirage 2000Ds flying with three 500lb Mk 82 unguided bombs under the fuselage (the fourth fuselage lateral hardpoint being usually occupied by a targeting pod). The Mk 82s are fitted with a DSU-33 all-weather, radar-ranging proximity sensor used to detonate the general-purpose warheads at a fixed height above ground to inflict the maximum amount of damage to scattered enemy forces. Like all Mirage 2000 two-seat variants, the 2000D is not fitted with internal guns.

The Mirage 2000D became the first tactical fighter to be armed with conventional cruise missiles when it was cleared to fly with a single Scalp under its centreline pylon. The 2000D community developed tactics and techniques to be exploited when firing Scalps to destroy pinpoint targets at stand-off distances with clinical accuracy. This missile that belongs to the Scalp / Storm Shadow / Black Shaheen family developed by MBDA has been successfully expended in combat by the Mirage 2000D and Rafale communities in Libya and in the Middle East, against Daesh.

Since the withdrawal of the Mirage F1CR, in 2014, the Mirage 2000D force has taken over the electronic intelligence role with the Astac (Analyseur de Signaux TACtiques, tactical signals analyser) pod previously flown by the F1CR recce strike fighter. Compared with more complex and heavier stand-off systems fitted, for example, to RC-135 ELINT aircraft, the Astac / Mirage 2000D pair can be utilised to penetrate hostile air-defences, forcing enemy operators to switch to their 'war frequencies': they do not know if the fighter is equipped with an Astac or with weapons, and have to react accordingly. Additionally, the Mirage 2000D boasts significant self-defence capabilities, and, if intercepted, can aggressively manoeuvre and even fire back to avoid destruction.

The Mirage 2000D is now being upgraded to the Rénovation à Mi-Vie (RMV) standard, meaning mid-life upgrade. It had earlier benefitted from the amelioration of various systems and from the introduction of the Link 16 datalink. Under the latest plans, a total of 55 Mirage 2000D fighters are expected to be brought up to RMV standard. The RMV name is a little deceiving as the aircraft receive only a limited upgrade that will allow them to continue for a few more years. Due to budget constraints, the scope of the RMV programme has been considerably scaled down. Initially, it had been envisioned to replace the Antilope terrain-following radar with the RDY-3 multimode radar, a development of the RDY that equips the Mirage 2000-5F, but this option was eventually rejected on cost grounds. As part of the RMV, the Mirage 2000D is receiving the Mica IR air-to-air missile that will supplant the Magic 2 short-range infrared-guided missile that is approaching the end of its useful life. It is also armed with a CC422 30mm compact gun pod that can be fitted under the left forward fuselage hardpoint, under the air intake.

Four squadrons are currently equipped with the Mirage 2000: Escadrons de Chasse 1/3 'Navarre', 2/3 'Champagne' and 3/3 'Ardennes', all stationed in Base Aérienne 133 Nancy-Ochey, and Escadron de Chasse et d'Expérimentation 1/30 'Côte d'Argent', at Base Aérienne 118 'Mont-de-Marsan'.

Right: The Mirage 2000D is currently undergoing a limited modernisation programme that will allow it to soldier on for at least another decade, probably more.

Below: The Mirage 2000D and the Rafale represent the French Air and Space Force's strike assets.

A Mirage 2000D armed with two GBU-49 Enhanced Paveway precision weapons photographed at Niamey, in Niger.

Two Mirage 2000D strike fighters await their turn to be refuelled.

A Scalp stealth cruise is clearly visible under this Mirage 2000D's centreline pylon.

A Mirage 2000D four-ship during a training mission. The drop tanks of the lead aircraft are painted in the markings of the 'Malzeville' Mirage 2000D maintenance squadron.

A Mirage 2000D sits in a shelter at Nancy. Mirage 2000D aircrews undertake a large percentage of their missions at night.

Mirage 2000D no 638 taxies back in at the end of a training sortie at Nancy-Ochey.

This tiger-striped Mirage 2000D is used from Mont-de-Marsan by Escadron de Chasse et d'Expérimentation 1/30 'Côte d'Argent' for various operational evaluation tasks.

Close-up on a Scalp stealth cruise missile under the centreline pylon of a Mirage 2000D. The blue line shows it is a training round.

Dassault Aviation Rafale C

The Rafale was conceived as the successor to the Mirage 2000 family. The main goal of the programme was to produce a fighter that would be capable of performing a whole range of missions with a single airframe, from both land bases and aircraft carriers. The plane needed to be significantly bigger than its predecessor to offer a much longer range and a considerably increased weapon load. As a result, the new design was to be powered by two engines, a major departure from the single-engine Mirage fighter family (Mirage III, Mirage 5, Mirage F1 and Mirage 2000) that had proved both highly successful in French and in foreign service and very appealing on the export market due to low acquisition costs when compared to bigger American jets such as the F-4 Phantom or F-15 Eagle. The Rafale would become only the second twin-engine supersonic combat aircraft to be produced by Dassault Aviation after the legendary Mirage IV imagined in the mid-1950s as the first French nuclear bomber. For the Rafale, Snecma (now Safran) specifically developed the M88-2 turbofan rated at 5,000kg (11,023lbs) dry and 7,500kg (16,535lbs) with afterburner. The M88-2 has proved to be a responsive, powerful and compact powerplant that has garnered respect and plaudits from pilots and maintainers alike for its performance levels, its reliability and its ease of maintenance. The Rafale also had to be extremely agile to prevail in the air-to-air arena, hence the advanced delta airframe with short-coupled canard foreplanes. Three variants were conceived: single-seat Rafale C (Chasse, or fighter) for the Air Force, single-seat Rafale M (Marine) for the Navy, and two-seat Rafale B (Biplace, or two seat). The Rafale N (Naval), a two-seat variant once envisioned for the Navy, was eventually cancelled, the Marine Nationale coming to the conclusion that the Rafale M single-seater could easily carry out even the most complex missions in the most demanding environments.

The Rafale has been conceived as an easily upgradable combat platform with a lot of growth potential. Since the first Rafale was delivered to the Armée de l'Air, in 2004, numerous successive standards have been introduced, culminating in the current F3R Standard. With each successive standards, new capabilities and weapons are adopted: air-to-ground weapons on Standard F2, nuclear missile, anti-ship missile, in-flight refuelling pod, Damoclés targeting pod and Pod Reco NG (new generation recce pod) on Standard F3, Meteor missile and Talios targeting pod on Standard F3R. On average, new software is introduced every two years. Nearly all the aircraft systems have been upgraded since service entry, including the RBE2 (Radar à Balayage Électronique 2 plans, or two axis electronic scanning radar that started its life as a PESA (Passive Electronically Scanned Array) radar but was fitted with a new state-of-the-art antenna of AESA (Active Electronically Scanned Array) technology in 2012, the Rafale thus becoming the first European aircraft equipped with an AESA. In early 2023, the French armed forces started switching to the even more capable Standard F4.1 optimised for better connectivity as part of the digital battlefield concept. It will be followed by F4.2 in 2025.

The Rafale can carry a wide array of precision weapons, with an external payload of 9,500kg (20,944lbs) spread under 14 hardpoints (13 'only' on the Rafale M naval variant). It is armed with a single internal 30mm 30M791 cannon, a powerful weapon designed to fire 21 rounds in half second bursts. In all, 125 rounds are carried. For air-to-air combat, it is armed with the Mica IR and RF. Since the Meteor's entry into service with the French Air and Space Force and French Navy Rafales, the Mica is no longer the longest-range air-to-air missile in the French inventory. Derived from the Mide (Missile d'Interception à Domaine Elargi, missile with an enlarged interception envelope) requirement, the Meteor is a European collaborative programme under British leadership. It was not conceived as a replacement for the Mica, however, but rather as a higher end answer to a need for a much longer-range missile propelled by a ramjet to defeat targets at extreme ranges. The end of the Cold War and the ensuing reduction in defence budgets in Europe led to a protracted development programme. Meteor service entry on the Rafale began only in 2019.

For the air-to-surface missions, the Rafale carries laser-guided bombs of the Paveway series, including the GBU-12, the GBU-16, the GBU-22 and the GBU-24, and French-designed precision weapons of the Hammer (Highly Agile, Modular Munition Extended Range) series. For precision attacks of heavily defended targets deep inside enemy territory, the Scalp stealth cruise missile is the Rafale's weapon of choice. Up to three Scalps can be carried under the fighter's wings and fuselage, but two under the wings are a more usual load. The range of the Scalp missile is said to be about 400km (250 miles).

At the time of writing, 79 Rafale C single-seaters had been ordered by the AAE, but 20 had been exported as second-hand airframes to Greece and Croatia after the two countries had selected the Rafale. Contract for another batch of Rafales was awaited when this book was being written. Another airframe, Rafale C101, has spent all its life at Istres with the Dassault Aviation Flight Test Centre, serving as an instrumented test bed for the development of future software drops and standards.

Rafale C single-seaters are allocated to Escadron de Transformation Rafale 3/4 'Aquitaine', the Rafale operational conversion unit at Base Aérienne 113 Saint-Dizier, and to Régiment de Chasse 2/30 'Normandie-Niémen' and Escadron de Chasse 3/30 'Lorraine', both stationed in Base Aérienne 118 Mont-de-Marsan. Escadron de Chasse et d'Expérimentation 1/30 'Côte d'Argent' also fields a limited number of Rafales for operational evaluations from Mont-de-Marsan. Finally, Escadron de Chasse 1/7 'Provence' operates six Rafale C fighters from Base Aérienne 104 Al Dhafra, in the United Arab Emirates.

This Rafale is armed with a Scalp cruise missile under the centreline pylon and two Mica IR air-to-air missiles at the wing tips.

The Rafale is a powerful beast. This aircraft flying in the vertical is fitted with two Smokewinders' smoke generators at the wing tips for airshow displays.

Four Rafale C fighters share the ramp during an exercise in Solenzara, Corsica.

During the war in Libya in 2011, this Rafale C takes off from Solenzara, in Corsica. It is armed with four Mica RF/IR missiles and equipped with two 2,000-litre drop tanks and a Pod Reco NG recce pod under the centreline pylon.

Dassault Aviation Rafale B

The Rafale B two-seat variant of the Dassault Aviation fighter is in service both for combat missions and in the conversion training role, to help new pilots straight from training transition on to the Rafale.

Two squadrons of the Forces Aériennes Stratégiques, the French strategic air command, fly the Rafale B in the nuclear deterrence role from Base Aérienne 113 Saint-Dizier. They are armed with the ASMP-A (Air-Sol Moyenne Portée Amélioré), improved medium range air-to ground nuclear-tipped missile that entered service in 2009. The ASMP-A is powered by a ramjet engine at speeds said to exceed Mach 3 at high level and Mach 2 at low level, although no reliable figures have officially been confirmed. Range is said to be twice that of the earlier ASMP missiles that equipped the Mirage 2000N and the Super Étendard Modernisé. An even more capable version of the ASMP-A was qualified in early 2022. However, nothing has been released regarding the improvements introduced, but they are thought to focus on the navigation suite to further improve the precision of the missile. While Navy Rafale M fighters are flown by a single pilot, even for nuclear missions, the French Air Force has decided that the Rafales selected for the nuclear deterrence role will be two-seaters in order to allow the crew to use a 'double key' to authorise the launch of the missile.

The Rafale was designed from the outset with radar and digital terrain matching / automatic terrain-following systems to hide from ground-based surveillance radars and surface-to-air missile systems, and with advanced navigation systems for autonomous and tremendously accurate navigation over very long distances, even when outside signal is not available from TACAN, VOR, DME beacons and from GPS and Galileo satellites. The two FAS squadrons are the specialists of low-level, high-speed missions, exploiting to the full the Rafale's automatic terrain-following capabilities to avoid detection and relying on its air-to-air sensors (radar and infrared search and track) and its Mica and Meteor missiles to destroy enemy fighters that might try to intercept them. FAS aircrews can also perform conventional missions, and they regularly deploy to the Middle East and to Africa to share the burden with the Mirage 2000 and Rafale C communities. They also participate in the homeland defence mission alongside Mirage 2000-5F interceptors and single-seat Rafale C fighters.

The Rafale B is fitted with the same avionics suite as the Rafale C and M variants, and can carry the same conventional weapons under the same number of hardpoints as the Rafale C (14). Like the single-seat Rafale C and Rafale M variants, the Rafale B is armed with a single 30mm 30M791 gun with 125 rounds. However, its internal fuel capacity is reduced from 4,700kg (10,362lbs) to 4,400kg (9,700lbs) to make room for the rear cockpit and its second Martin-Baker Mk 16F zero-zero ejection seat (produced under licence in France).

At the time of writing, 65 Rafale B two-seaters had been ordered by the AAE, but four had been sold as second-hand airframes to Greece and Croatia. Two heavily modified aircraft, the B301 and B302, are permanently assigned to Istres where they undertake various trials with the Dassault Aviation Flight Test Centre. Like the Rafale C single-seaters, all Rafale B fighters are regularly upgraded to ensure a high-level of commonality between the two variants. At the time of printing, all Rafale Bs had been brought up to F3R Standard.

The two aforementioned FAS squadrons, Escadrons de Chasse 1/4 'Gascogne' and 2/4 'La Fayette' are both stationed in Base Aérienne 113 Saint-Dizier. Co-located with them at Saint-Dizier is Escadron de Transformation Rafale 3/4 'Aquitaine', the Rafale joint operational conversion unit that trains all future French Rafale pilots (Air Force and Navy) and some foreign pilots as part of agreements following the sale of Rafales to export customers. At Base Aérienne 118 Mont-de-Marsan, Escadron de Chasse et d'Expérimentation 1/30 'Côte d'Argent', the French operational evaluation unit also flies a limited number of two-seat Rafales.

A Rafale in playful mood. The Rafale is a notoriously powerful and agile fighter.

Rafale in a typical training configuration, with two 2,000-litre drop tanks. The two-seat Rafale B variant is fully capable of conducting the whole range of missions the fighter was designed for.

This nuclear-capable Rafale B taxies out for yet another sortie during Maple Flag 2016, at Cold Lake.

At the request of the author of this book, the pilot of this Rafale shows how the landing gear retracts while flying at medium altitude.

A Rafale B initiates a strafing run at the Captieux range.

This Rafale B taxies out of its hardened aircraft shelter at Saint-Dizier. HASs are very effective at protecting fighters against light UAVs.

During Maple Flag 2016, a Rafale B armed with four GBU-12 laser-guided bombs overflies the large expanses of the Cold Lake range, in Canada.

The Rafale B specialises in the training and nuclear deterrence roles but is also engaged in other missions, including precision strikes, close air support and air defence.

Two Rafale B squadrons specialise in the nuclear deterrence role. This aircraft belongs to Escadron de Chasse 'Gascogne' stationed in Saint-Dizier.

Chapter 2
Command and Control / Intelligence-Collecting Aircraft

Boeing E-3F Sentry

The French Air and Space Force operates a fleet of four E-3F Sentry aircraft stationed in Base Aérienne 702 Avord, in the centre of the country. The French Air Force procurement programme was launched in the '80s when a need to detect and track extremely low-flying intruders such as Soviet Su-24 Fencers emerged in France. At the time, the US Air Force and NATO were busy introducing into service the E-3 Sentry, widely known as the AWACS (Airborne Warning and Control System) and that aircraft was the obvious choice for the French Air Force. Other options were considered, including the E-2C Hawkeye and the Nimrod AEW3 then being developed for the Royal Air Force. Eventually, the E-3F Sentry was chosen and four were delivered in 1990/1991, officially entering front-line service in 1992. They are operated by the 36ème Escadron de Détection et de Contrôle Aéroportés 'Berry' and regularly co-operate with US Air Force and NATO AWACSs and with the French Navy E-2C Hawkeyes of Flottille 4F.

Like the British E-3D and Saudi E-3S variants, French Sentries were powered by fuel-efficient CFM56 turbofans from the beginning. They were fitted with both a refuelling probe above the right side of the forward fuselage for probe and drogue refuelling from Royal Air Force Tristars and VC-10s, and with a boom receptacle above the cockpit for boom-equipped tankers such as the C-135FR/KC-135R Stratotanker and the KC-10 Extender. The E-3F probe was withdrawn, however, in the late 2000s to cut down maintenance requirements and fuel consumption, with the added benefit of a massive reduction in aerodynamic noise levels in the cockpit.

The E-3F is the main French Air Force airborne command and control asset. The E-3F also has a secondary electronic intelligence-gathering mission, helping determine hostile nations' electronic order of battle and update French electronic warfare threat libraries that will, in turn, be loaded into combat aircraft's electronic self-defence suites.

The four Sentries were extensively modernised as part of a programme launched from 2014. The in-depth upgrade encompassed the replacement of the mission calculator and of the ten work stations, the adoption of a further four work stations to increase the airborne command and control capabilities, and of a new IFF system to bolster the interoperability with NATO allies. The four French aircraft were modified during depot-level maintenance periods from 2018 to 2020. As a result, they are now similar to the US Air Force's E-3G Sentry Block 40/45, ensuring full interoperability with US and NATO Sentries. In 2017, it was announced that the E-3F Sentry's connectivity would be further improved via the adoption of a SATCOM (Satellite Communication) datalink and that the cockpit would be significantly upgraded thanks to the introduction of five multifunction displays instead of traditional 'steam gauges'. The adoption of the new cockpit will lead to a reduction of the flight deck crew from four to three aircrews. A final modernisation programme was announced in July 2022. This will allow the E-3F to continue in service until 2035, at least.

An E-3F Sentry photographed during a ceremony at Avord, in June 2022. Noteworthy is the dismounted in-flight refuelling probe.

Undercarriage and flaps fully down, this E-3F Sentry undertakes a slow pass during an airshow.

Beech King Air 350 VADOR

The Beech King Air 350 VADOR (Vecteur Aéroporté de Désignation, d'Observation et de Reconnaissance), or airborne designation, observation and reconnaissance vector is one of the most recent types to have entered service in France. Selected as part of the Avion Léger de Surveillance et de Reconnaissance (ALSR, or light surveillance and reconnaissance aircraft) programme, the VADOR has been adopted to start replacing aircraft on loan from the private industry, such as the Merlin IV twins from CAE Aviation used for a wide range of roles in France and abroad. Like the contracted aircraft, the VADOR is designed to perform a wide range of Intelligence, Surveillance, Target Acquisition and Reconnaissance (ISTAR) missions and offers capabilities broadly similar to those of the US Air Force's MC-12W Liberty or to the Royal Air Force's Shadow R1/R2.

Operated by a crew of five, composed of two pilots, a tactical coordinator, a communication intelligence operator and an image intelligence operator, the VADOR is fitted with a comprehensive multi-spectral sensor suite that is known to include a belly-mounted electro-optical turret for 360-degree surveillance. Besides, it offers significant signal intelligence-gathering capabilities.

It is flown from Base Aérienne 105 Évreux-Fauville by Escadron Électronique Aéroporté 1/54 'Dunkerque'. At the time of writing, two had been delivered, with a third one to follow by 2025. The overall requirement is stated at eight VADORs, but is likely to be cut down to six due to evolving needs. The vulnerability of such aircraft for high-intensity warfare scenarios is certainly a cause of concern for the AAE, and the usefulness of the VADOR is more or less limited to counter-terrorism operations or surveillance missions in a permissive environment. This probably explains the likely decision to reduce the number of aircraft to be purchased at a time when international tensions are on the rise following the invasion of Ukraine by Russia.

The new VADOR has already played a crucial role providing airborne surveillance capabilities in the fight against terrorism.

Like many other countries, France has selected the King Air, a proven family of turboprop aircraft, as their next manned surveillance asset.

It is highly likely that the VADOR will be drafted in to help secure the Paris 2024 Olympic Games.

General Atomics MQ-9 Reaper

Unmanned Aerial Vehicles (UAV) are progressively taking a central role in most modern air forces and the Armée de l'Air et de l'Espace is no exception. The service has long operated UAVs for limited reconnaissance and surveillance missions, but its drone force really became operational with the successive introductions of the IAI/EADS Harfang and of the General Atomics MQ-9 Reaper. The piston-powered Harfang was the first long-endurance UAV operated in France and it helped build up experience and refine operational concepts relying on satellite links for very-long range missions deep inside 'bandit country' and exploiting to the full its sensor suite. After a successful operational career, including combat operations in Mali, it was withdrawn from use in January 2018, leaving the more capable General Atomics MQ-9 as the main AAE UAV.

The MQ-9 Reaper was ordered in 2013 as part of an urgent operational requirement to significantly increase French surveillance capabilities in Africa. Initially, four systems, each with three Reapers, were delivered to the Armée de l'Air. Since entering service, the Reapers have earned an enviable operational reputation, helping disseminate invaluable intelligence to the warfighters on the ground. As a result, they are in high demand. The Reaper is equipped with a comprehensive sensor suite composed of the Lynx radar and of the Raytheon MTS-B (Multi-Spectral Targeting System-type B) electro-optical turret optimized for high-level operations. The MTS-B turret offers startling performance levels. Compared to that of the Harfang, its detection and identifications ranges are considerably longer while flying at a much higher altitude. It is fitted with a wide array of internal systems:

TV / infrared / near infrared / low light level TV multi-spectral sensors.
laser illuminator for the guidance of precision munitions.
laser pointer.
laser rangefinder.

French MQ-9 Reapers have proved extremely useful fighting terror groups in the vast expanses of the Sahara.

This head-on shot of the MQ-9 Reaper advantageously shows the configuration of its tail surfaces.

The Harfang was a much older design that did not have the latest sensors, but it proved invaluable to develop operational procedures.

The Harfang was used by France to gain operational experience in the use of long-range UAVs.

Command and Control / Intelligence-Collecting Aircraft

The Reaper is remotely piloted from dedicated control stations that can be hidden from hostile eyes.

The French Air and Space Force has all the necessary mobile command and control shelters to operate long-range UAVs from forward operating bases.

The Lynx radar offers new capabilities to the French aviators. With its high-resolution Synthetic Aperture Radar (SAR) and Ground Moving Target Indicator (GMTI) modes, it has become a key surveillance tool to monitor movements in the deep desert. The system has proven highly flexible to operate and the operator can change from one mode to another at the flick of a switch. The SAR mode proves ideal to obtain extremely accurate radar imagery at extreme range, day and night, even through clouds. The SAR operates either in spot mode, to monitor a given area, or in strip mode, to scan a long swathe of desert, along a dirt track for example. As its name suggests, the GMTI is used to detect and track vehicles moving around. The MTS-B turret can be instantly pointed at a target of interest detected by the radar.

The MQ-9's ground control station – or cockpit – accommodates a crew of two, pilot and sensor operator. The two crew members have, at their disposal, no fewer than 14 screens, two of them shared by the pilot and the Senso. The pilot is obviously in charge of flying the Reaper, but he also handles the Lynx radar. Another two personnel – an intelligence officer, acting as tactical coordinator, and a photo interpreter in charge of the real-time imagery processing – assist the two aircrews. The photo interpreter plays a decisive role capturing, editing, processing and forwarding to the local commander all imagery of interest: it is a very quick process that provides the most up-to-date data to the intelligence cell, allowing crucial information to be relayed to the warfighter in the field in the shortest amount of time in order to enhance

The Reaper is designed to be easily serviced and maintained in bare forward bases.

battlefield awareness. The Reaper's communication suite also permits the crew to quickly call for a lethal array of supporting fire for troops in contact, from artillery to Rafales and Mirage 2000 fighters.

With the introduction of the Block 1 variant, in 2019, the AAE now has at its disposal an armed variant of the Reaper that can deliver kinetic effects from the air, thus providing within one single airframe both surveillance and close-air-support/attack capabilities for maximum combat efficiency. Two Block 1 systems with six Reapers were delivered before production switched to the improved Block 5 (another two systems, with six UAVs). French Block 1 and 5 Reapers are mainly armed with GBU-12 Paveway II laser-guided bombs. All Block 1 Reapers will eventually be upgraded to Block 5 standard.

Moreover, the Reaper can be used for surveillance missions as part of the wide homeland defence mission. It can take part in the protection of major events such as the G8 or G20 summits or the D-Day anniversary celebrations. It is highly likely that Reapers will help secure the Paris Olympic Games, to be held in 2024.

Two front-line squadrons, Escadrons de Drones 1/33 'Belfort' and 2/33 'Savoie', operate the MQ-9 from Base Aérienne 709 Cognac-Châteaubernard. They are supported by a conversion unit, Escadron de Transformation Opérationnelle Drone 3/33 'Moselle', also stationed in Cognac. Only a limited number of drones are flown from Cognac, however, with most Reapers operating from Niamey, Niger, one of the main French forward operating bases in Africa.

The Reaper's sensor suite encompasses powerful day and night cameras that can detect pinpoint targets at very long ranges.

Chapter 3
Tankers and Airlifters

Boeing C-135FR Stratotanker

Since the 1960s, the C-135FR has been the workhorse of the French tanker force. The initial C-135F was basically a KC-135A with some features borrowed from the USAF's C-135B cargo aircraft. As such, it was a kind of Multi-Role Tanker Transport (MRTT) well in advance of its time. Its main cabin can be arranged in various ways to accommodate passengers (a maximum of 126 persons), pallets (nine, weighing up to 3,600kg /7,900lb each), stretchers (up to 40), or a combination of any of these loads. The C-135F initially focused on refuelling the Mirage IV nuclear bombers as part of the French nuclear force, but its missions were progressively expanded. For example, it supported French nuclear trials in the Pacific, conducting long-range weather reconnaissance missions and nuclear air-sampling sorties.

Out of the 12 C-135Fs originally delivered by Boeing from 1964, one was lost in June 1972, in a fatal crash, in Hao Atoll, in French Polynesia. The 11 remaining aircraft were all converted to C-135FR standard between 1985 and 1988. This comprehensive upgrade programme involved the replacement of the thirsty Pratt & Whitney J57-9-59W turbojets with water injection by four CFM International CFM56-2B1 high bypass-ratio turbofans (military designation F108-CF-100) rated at 97.86kN (22,000lb). With the CFM56, total thrust was augmented by 40–50 per cent, depending on the temperature and pressure conditions, and fuel consumption was reduced by 27 per cent on average, helping increase the give-away capability. From 1989, the C-135FRs were all fitted with a new autopilot derived from that of the Atlantique 2 maritime patrol aircraft, together with a new instrument panel broadly equivalent to that of the Airbus A300 airliner. In mid-1993, the fleet started receiving Flight Refuelling Ltd Mk 32 refuelling pods mounted under the wings, considerably augmenting flexibility and safety. The first C-135FR was modified in Wichita, Kansas, while the remaining ten were fitted with the pods after undergoing an upgrade programme by Air France Industries, in Orly airport, near Paris. Simultaneously, the tankers' fuel system was modified with more powerful pumps that provide increased transfer rates, a change needed to refill the E-3F AWACS that had just entered service at the time. The modification was required to accelerate the refuelling of the larger aircraft. It is worth noting that a significant percentage of the C-135FR/KC-135RG boom operators are trained to operate the rigid boom refuelling system to refill the E-3F AWACSs or foreign boom-compatible aircraft.

The C-135FR is flown by a crew of four, two pilots, one navigator and one boom operator. In the strategic role, the C-135FR can be utilised as a C^2 (Command and Control) aircraft, allowing the French strategic air command decision makers to stay in touch with the Rafales, even at long range thanks to the tanker's HF radios and to a recently introduced SATCOM (Satellite Communication) system.

Medical evacuation has become a major mission for the C-135FR fleet. The terrorist attack against French defence contractors in Karachi, in Pakistan, in 2002, and against the French military base in Bouaké, in Ivory Coast, in 2004, had shown a shortcoming in massive medical evacuation capabilities in France, prompting the development of the Morphée (Module de réanimation pour Patients à Haute Élongation d'Évacuation, or long-range patient evacuation and resuscitation module) for the Stratotanker. The Morphée kit allows up to 12 patients to be evacuated in one go. Two configurations are available: one version with six intensive care beds for critically wounded patients imposing the use of oxygen and of advanced monitoring systems, and a second, more flexible variant with four ICU beds and the associated equipment, and another eight beds for less severely hurt persons.

As part of the 'Réno 2' programme, the avionics suite of the 11 remaining C-135FR Stratotankers recently underwent a last round of modernisation with the introduction of a modern weather radar with a larger and clearer screen, of a new fuel-management panel with a digital display, and with modern Flight Management Systems for the main instrument panel and for the navigator's workstation.

In French service, the Stratotanker has proved to be a tough and reliable aircraft, but the type is becoming increasingly difficult and expensive to maintain, requiring a fast-mounting number of maintenance hours per flying hour. Thankfully, its successor, the Airbus A330 MRTT Phénix is there in growing numbers. The C-135FR will be flown until final retirement by Escadron de Ravitaillement en Vol 4/31 'Sologne', at Base Aérienne 125 Istres-Le Tubé. The unit was recreated as a temporary squadron when the 'Bretagne' name was re-allocated to the first Phénix unit. 'Bretagne' is an extremely prestigious name, one of five Free French squadrons of World War Two fame, alongside 'Ile-de-France', 'Alsace', 'Lorraine' and 'Normandie-Niémen', explaining why it was preserved early. The first C-135FR, s/n 475, was withdrawn from use in October 2020 and another six had followed at the time of writing, one aircraft retiring for each A330 Phénix delivered.

This C-135FR Stratotanker takes off from Mont-de-Marsan air base in April 2017.

N'Djamena has long been a forward operating base for the C-135FR. This aircraft is photographed in Chad in 2015.

Boeing KC-135RG Stratotanker

To supplement the 11 C-135FR tankers still in service at the time, a further three KC-135R tankers were obtained on loan from the USAF in 1992 to expand the French tanker fleet. The operations in Iraq (Desert Shield and Desert Storm) and in Bosnia, and the increasing number of French fighters fitted with a refuelling probe had helped confirm the need for more tankers to expand long-range combat capabilities.

Delivered as part of the 'Peace Armagnac' programme, the three KC-135Rs were eventually purchased from the USAF in 1997. Unfortunately, they were not up to the same standard as the original French aircraft: they have a wooden cargo floor instead of the steel version that equips the C-135FR; they are fitted with a single air-conditioning system instead of two on the FR variant; and their navigation system was not as modern. Furthermore, they are not equipped with a liquid oxygen system, relying instead on a gaseous oxygen supply. This is why the Armée de l'Air decided against fitting them with the Mk 32 in-flight refuelling pod. As a consequence, they were initially mainly tasked for training, leaving the C-135FRs to concentrate on operational missions.

Under a contract signed with Rockwell Collins in 2013, the three KC-135R tankers were upgraded to KC-135RG Block 40.5 GATM (Global Air Traffic Management) standard, with the final aircraft delivered back from the San Antonio plant (Texas) to France in 2015. The upgrade brought the three of them to a standard similar to those updated for the USAF, thus improving interoperability while ensuring a high level of commonality between US and French Stratotankers, a major advantage that helped reduce costs. They are now fitted with an Inmarsat 4 SATCOM (Satellite Communication) system, secure voice encrypted radios, and the Aircraft Communications Addressing and Reporting System (ACARS), a civilian digital datalink used for long-range air-traffic control. The modernised KC-135RG is flown by a crew of three, two pilots and a boom operator. Full operational capability of the RG standard was approved in 2015 after a short operational evaluation.

The C-135FRs had logged considerably more flying hours than the three ex-USAF aircraft. This is the reason why the three KC-135RG Stratotankers will remain in service until 2027 within Escadron de Ravitaillement en Vol 4/31 'Sologne', at Base Aérienne 125 Istres-Le Tubé. The decision made in the early 90s to procure these three additional Stratotankers has proved to be extremely wise, giving the opportunity to boost the French Air Force's operational capabilities at an acceptable cost. When they are eventually withdrawn from use, in 2027 under the latest plans, they will have logged thousands of flying hours over a career, in French colours, spanning more than 30 years.

A neat row of C-135FR and KC-135RG tankers at Istres. All the French tankers are operated by the Forces Aériennes Stratégiques, the French strategic air command that handles the nuclear deterrence mission.

A fully armed QRA Rafale slowly approaches a KC-135RG. French tankers take an active part supporting the homeland defence.

This Swiss Air Force F/A-18C Hornet has just been refuelled by a KC-135R over eastern France.

With its wooden cargo floor, the KC-135RG is less useful than the C-135FR in the logistics transport role.

Airbus A330 MRTT Phénix / Airbus A330 AUG

Replacing the Stratotanker fleet has long been a priority for the French MoD, but budget constraints have derailed the programme on numerous occasions. At some stage, France considered buying the Airbus A310 MRTT (Multi-Role Transport Tanker) that had been selected by Germany and Canada, but that project did not materialise. The launch of the A330 MRTT programme by Airbus was welcome news for the Armée de l'Air and the aircraft was selected as the obvious replacement for the Stratotanker, with a global need for 15 airframes to supplant the C-135FR/KC-135RG, the A310 and the A340.

In comparison to the Stratotanker, the Phénix offers considerably expanded operational capabilities thanks to a much higher maximum weight and larger fuel capacities. For example, a single Phénix can refuel two fully-armed Rafales engaged in a 10,000km (6,210 miles) nuclear raid while a C-135FR is restricted to 8,000km (4,971 miles). This represents a 25 per cent increase in range. It also provides aircrews with a much better connectivity with the fighters and the higher echelons thanks to the L16 datalink and to a satellite communications system. In the force deployment role, the fleet of Phénix MRTTs will eventually be large enough to project a squadron of 20 Rafale fighters and all its support equipment over a distance of 20,000km (12,420 miles) in less than 48 hours, thus allowing France to rapidly reinforce its military forces in French overseas territories in the Pacific.

Since entering service, the Phénix has proved ideal for the long-range passenger / cargo transport role, thus demonstrating in real conditions that it can advantageously supplant the now retired A310s and A340s that have faithfully served the French Armed Forces. While they cannot accommodate any large vehicles, their cargo holds are big enough to carry large items such as ammunition containers, combat dogs in special air transportable kennels, dedicated crates for targeting pods, or spare M88-2 turbofan engines for the Rafales. All French MRTTs are equipped with 88 passenger seats at the rear of the main cabin, but they can be fitted with an additional 184 seats when required.

The Phénix is also progressively taking over the medical evacuation mission from the C-135FR fleet. The Morphée kits developed for the C-135FR have been adopted for the Phénix too. In 2020, at the peak of the Covid crisis, the A330 MRTT fleet was drafted in for medical evacuations of critically ill patients from areas where hospitals were experiencing difficulties, transferring a growing number of patients to areas where the pressure on hospitals was less severe. Technical solutions were quickly developed by CBRN (Chemical, Bacteriological, Radiological and Nuclear) experts to isolate the cockpit from the main cabin and to completely disinfect the aircraft at the end of each mission.

The A330 MRTT has been operated by Escadron de Ravitaillement en Vol et de Transport Stratégique 1/31 'Bretagne' (ERVTS, or in-flight refuelling and strategic transport squadron) at Base Aérienne 125 Istres-Le Tubé since October 2019. All Phénix aircrew training is undertaken within Escadron de Transformation Phénix 3/31 'Landes' (ETP 3/31). The Phénix is usually flown by a crew of four (two pilots, an in-flight refuelling operator and a flight attendant for the safety of the main cabin), but additional crew members are sometimes carried for complex and / or long-duration missions. All these squadrons belong to the 31ème Escadre Aérienne de Ravitaillement et de Transport Stratégique, the 31st in-flight refuelling and strategic transport wing. A second front-line squadron, ERVTS 2/31 'Esterel', will convert to the Phénix in 2023. By then ERVTS 1/31 and 2/31 will be organised along the same lines, with the same missions and the same number of aircrews who will share the pooled Phénix fleet with ETP 3/31.

From 2025, the A330 MRTT fleet will be equipped with broadband SATCOM terminals that will be fully compatible with Syracuse 4 satellites (the latest generation of French military satellites), satellites from allied NATO nations and with civilian satellite constellations, thus considerably improving the C^2 capabilities of the tankers. The Ka-band SATCOM will allow tanker aircrews – and accompanying fighters – to remain at all times in constant contact with higher echelons, a major advantage for long-range missions, especially as part of the French nuclear deterrence concept.

A single A330-223 AUG (Avion à Usage Gouvernemental, or governmental use aircraft) is in AAE service for long-range VIP transport missions. It entered service in 2010. It is configured to carry up to 76 passengers in very comfortable conditions and is equipped with a wide assortment of communication equipment enabling the Président de la République to remain in constant contact with higher French authorities.

In 2020, two A330-200s were acquired on the second-hand market as part of a plan to support the aviation industry during the Covid crisis. The two relatively low flight-hour airframes have been delivered in a standard passenger/airline configuration but will eventually be brought to MRTT standards.

Right: On top of its tanker role, the Phénix can be used as a transport aircraft for up to 272 passengers.

Below: A French Phénix shares the ramp with Royal Australian Air Force KC-30As at RAAF Amberley, during Pitch Black 2022.

Left: The refuelling probe of this Rafale C has just been plugged into the left basket of a Phénix tanker.

Below: Four Armée de l'Air et de l'Espace Phénix tankers photographed at their Istres home base.

Above: A Mirage 2000-5F interceptor is being refuelled by a Phénix tanker over Poland in April 2022.

Right: Initially developed for the C-135FR, the Morphée medical kit has also been adopted for the Phénix.

Below left: The Phénix is equipped with an advanced cockpit typical of all recent Airbus instrument panels.

Below right: The ARO's workstation is a sophisticated piece of equipment that allows the operator to maintain a careful watch of what is happening behind the tanker.

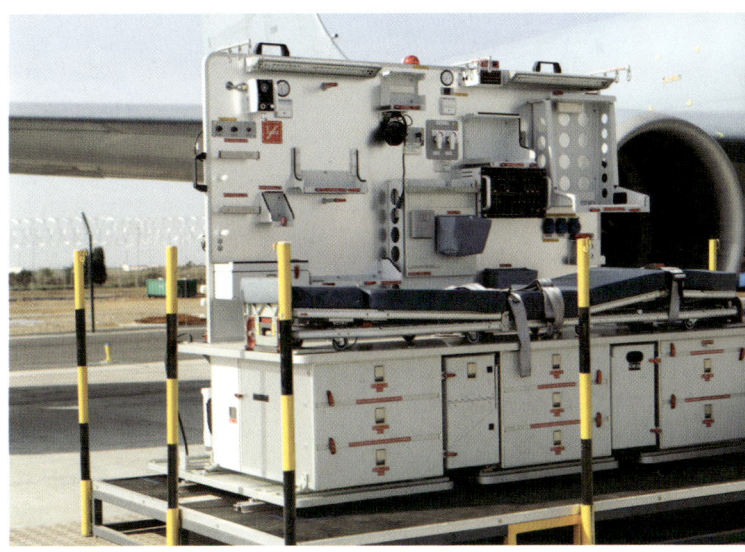

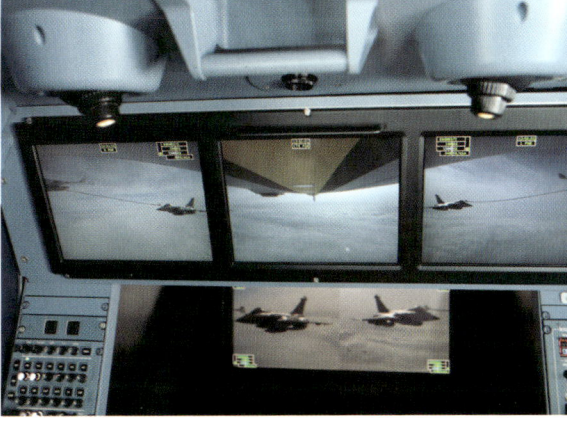

Airbus A400M Atlas

The A400M Atlas was conceived by Airbus as the next generation airlifter for European air arms. This collaborative programme had been launched to provide European countries with a common advanced design that would offer considerably improved transport capabilities when compared to the ageing Hercules and Transalls it was due to replace. After a protracted development and trials programme, the first production A400M was delivered to the Armée de l'Air in September 2013. As these lines were written, the AAE had received 19 aircraft out of 25 planned to be in service by 2025. The total requirement is still officially at 50, although it remains to be seen if these 50 aircraft will all be purchased.

Service entry has been difficult, with extremely bad initial availability rates and countless problems that led to huge maintenance requirements. A seemingly endless list of teething problems had to be addressed too, but the situation was slowly improving at the time of writing thanks to the efforts of the AAE air and ground crews, their multinational partners and the industry. The first aircraft lacked some of the systems and functionalities envisioned for the type: for example, the first few were incapable of dropping paratroopers, a major drawback for a tactical airlifter, but Atlases were delivered in progressively more and more capable standards that have benefited from the successive introductions of various subsystems. Many French pilots have already exceeded 2,000 flying hours on the type and experience is rapidly spreading among the aircrew and maintenance ranks alike.

From number 18, AAE A400Ms are now capable of conducting the whole spectrum of combat operations once envisioned for the Atlas. Numerous systems have been refined to expand their operational capabilities, improve reliability and facilitate maintenance. As a result, they will be able to operate more effectively in a contested airspace. They are equipped with the terrain-following system that will enable aircrews to fly below the enemy's radar coverage to undertake low-level penetrations. The severe technical limitations that restricted the number of paratroopers earlier aircraft could drop to 30 in one go have now been overcome, and since the delivery of the 18th aircraft in April 2021, the A400M has been able to deliver up to 116 paratroopers from its two rear side doors in one pass, a major operational step forward. Finally, the newest Atlases can be fitted with additional cargo hold tanks to increase the amount of fuel carried and boost their endurance and in-flight refuelling give-away capabilities. Earlier aircraft will progressively be upgraded to the newer standard.

While the AAE A400M force initially undertook unglamorous – but much needed – point A to point B logistics missions, the scope of the missions carried out by the Atlas is progressively being extended to tactical roles. The A400M has become the AAE's main tactical and strategic airlifter. Compared to the Transall it has replaced, it offers about twice the range and more than twice the payload. The Atlas is also considerably faster, thus providing much larger airlift capabilities than the now-retired twin. It is also far more capable and faster than both the C-130H Hercules and C-130J Super Hercules.

French A400Ms can now be equipped with pods to refuel fighters. Even though the AAE has purchased two KC-130J Super Hercules as gap fillers to refuel in flight its EC725 Caracal special forces / combat search and rescue helicopters, it is widely expected that the A400Ms will eventually also be used as tankers for helicopters as well as fast jets. Airbus test pilots have already conducted in-flight refuelling operations with Caracals during trials.

The AAE special forces community has already begun training on the Atlas and has tested the aircraft in a wide array of scenarios, including the air delivery of equipment and of large numbers of skydivers in one go, using HALO (High Altitude Low Opening) and HAHO (High Altitude High Opening) techniques from the rear ramp, relying on oxygen supply at such a high level. It is expected that special forces' aircrews will soon start routinely flying the Atlas in direct support of French special forces commandos, providing improved capabilities in terms of range and payload over the Hercules.

The French Air Force was the first military operator to receive the A400M, in 2013. At the time of writing, 20 A400Ms had been delivered to France.

Four French units, all stationed in Base Aérienne 123 Orléans-Bricy, now operate the Airbus A400M:

Équipe de Marque ATT, the French Air Warfare Centre project team that oversees all A400M operational evaluations.

Centre d'Entraînement des Équipages de Transport (CIET 340), the French Air and Space Force transport crew-training centre.

Escadron de Transport 1/61 'Tourraine'.

Escadron de Transport 4/61 'Béarn'.

ET 1/61 and 4/61 are now two sister units that share an increasing percentage of the tactical, pre-strategic and strategic airlift requirements of the French Armed Forces.

The French A400Ms have been successfully engaged in Africa, performing combat missions in support of French troops engaged against terror groups. In August 2021, they were drafted in to help evacuate French citizens and Afghan civilians stranded in Kabul, operating in difficult conditions alongside French Hercules and Allied assets.

The A400M is significantly larger and significantly more capable than the other two airlifters currently in service in France, the CN235 and the Hercules.

Lockheed Martin C-130H/H-30 Hercules

Fourteen C-130H/H-30 Hercules are in service with the AAE, performing a wide range of conventional and special forces' missions. The Hercules was purchased in 1984 by the French Air Force to supplement its Transall fleet and bring additional long-range airlift capabilities. The requirement had appeared after a number of operations, mainly in Africa, had shown shortcomings in terms of range and payload. The adoption of the classic Hercules was the only choice to address these shortcomings. Twelve were initially procured from Lockheed Martin, spread into six short-fuselage C-130H aircraft and six stretched C-130H-30 variants. The Hercules entered French Air Force service in 1987 and another two C-130H airlifters were subsequently purchased on the second-hand market to bolster the inventory and supplement the Transalls. Compared to the Transall, then the Armée de l'Air's main tactical airlifter and workhorse, the Hercules was heavier and more powerful and it offered longer range and better payload.

Ten C-130H/H-30 Hercules are in service with Escadron de Transport 2/61 'Franche-Comté' stationed in Base Aérienne 123 Orléans-Bricy. The unit also briefly operated the C-130J/KC-130J Super Hercules before the four new aircraft transferred to Base Aérienne 105 Évreux-Fauville (please refer to the following entry). The squadron undertakes both tactical and strategic airlift missions, in France and abroad. ET 2/61 is also due to transfer to Évreux in 2024.

Another two C-130H/H-30s are in service with Escadron de Transport 3/61 'Poitou', the dedicated AAE special operations airlift squadron also based in Orléans. They are fitted with additional long-range communication equipment to perform their missions. The two Hercules support special forces parachutists that rely on HALO and HAHO skydiving techniques for tactical insertion, either in daylight or under the cover of darkness, using oxygen. However, the aircraft is not considered as effective as the beloved Transall in terms of capabilities to operate from short, unprepared runways: the C-130 Hercules is renowned worldwide for its ability to land on unpaved runways thanks to its four large undercarriage main wheels, but few realise that the Transall offered even more impressive performance on rough dirt strips than its American competitor as it could operate from softer surfaces thanks to a very low ground pressure afforded by its eight main wheels. The AAE is currently pursuing a modernization programme for its special forces Hercules through the adoption of a MX-20 FLIR turret. They two will transfer to Évreux in 2024, along with their aircrews and engineers. They will still be operated by the Poitou for special missions, although the command structure will remain in Orléans.

A further two Hercules are known to be serving with Groupe Aéromobile 56 'Vaucluse' (GAM 56), the secretive squadron that operates in support of clandestine operations from Évreux-Fauville. The squadron is cleared to perform very low-level penetrations with Night Vision Goggles and to carry out air drops of special operators and cargo from low level to very high levels, relying on various techniques and on an assortment of parachutes, containers and platform types.

Collins Aerospace France has launched a modernisation programme of the AAE C-130H/H-30 fleet. All 14 Hercules are now being equipped with a new generation of glass cockpit that will replace outdated 'round dials'. The Collins Aerospace Flight 2 suite has been selected: among other refinements, the modernised C-130s are being fitted with seven NVG-compatible multifunction displays, with Head-Up Displays, with an infrared sensor and with the multi-spectral EVS-300 enhanced vision system that will all help pilots operate in bad weathers.

A short fuselage C-130H lifts off for yet another transport mission.

Above: The French C-130H/H-30 fleet is currently relocating from Orléans to Evreux, Orléans becoming an all-A400M base in the process.

Right: Special forces Hercules aircrews are cleared to fly at night at very low level with the help of night vision goggles.

Below: French Air and Space Force C-130H/H-30 Hercules are flown with underwing fuel tanks most of the time.

Lockheed Martin C-130J / KC-130J Super Hercules

The delays that affected the service entry of the A400M Atlas and its severe teething problems, the approaching withdrawal of the Transall fleet and the chronic availability problems of the ageing C-130H/H-30 fleet pushed the French Air Force to order two C-130J-30 Super Hercules transport aircraft and two KC-130J tanker variants to fill a growing operational gap. The programme quickly moved forward, with contract signature with Lockheed Martin in January 2016. The two long-fuselage C-130J-30s were delivered in December 2017 and June 2018 respectively, with the two KC-130Js following in September 2019 and February 2020.

In terms of operational capabilities, the C-130J-30 represents a significant leap forward compared to the earlier C-130H/H-30. It is more powerful, carries more, and flies higher, faster and further. It is flown by a crew of four (two pilots and two loadmasters) while the C-130H requires two additional crew members, navigator and flight engineer. Unlike the A400M, it cannot carry the latest French Army armoured wheeled vehicles, however, a shortcoming that has appeared with the introduction of a new generation of combat vehicles as part of the Armée de Terre's Scorpion programme.

The two short-fuselage KC-130J now regularly co-operate with AAE EC725 Caracal helicopters, thus supporting both special operations and combat search and rescue training. The KC-130J tanker variant can also be drafted in for the more traditional transport role. Although service entry of these four aircraft has been quite recent, they have already been engaged operationally in the Sahara and have proved their worth on numerous occasions. For example, they have actively participated in the recent withdrawal of French forces from Mali.

Although the type entered service at Base Aérienne 123 Orléans-Bricy, it has been decided that the four French Super Hercules will now be operated from Base Aérienne 105 Évreux-Fauville alongside six Luftwaffe Super Hercules (three stretched C-130J-30s and three KC-130J tankers). Once all-German aircraft are delivered by the end of 2023, they will form a totally integrated bi-national squadron to be named 'Rhin'. The ten-aircraft squadron will be a fully multi-role unit capable of undertaking a wide range of roles.

It was widely expected that some additional Super Hercules would be purchased to start replacing the C-130H fleet and / or bolster the current C-130J/KC-130J fleet, but studies have now been launched to define what could be a future European airlifter potentially called A200M. It would be modern day version of the Transall, with state-of-the-art technologies and excellent rough field performance, an absolute necessity in Africa where AAE aircraft routinely operate from unprepared dirt or laterite strips.

The C-130J-30 and KC-130J can easily be distinguished from AAE C-130H/H-30 by their six-bladed propellers.

While a Phénix takes off in the background, the crew of this KC-130J goes through the checks before taxiing for yet another mission.

Above: The short-fuselage KC-130J is fitted with two massive refuelling pods under the wings.

Right: The C-130J-30/KC-130J fleet has already proved its worth, especially in Africa.

De Havilland Canada DHC-6 Twin Otter

The DHC-6 Twin Otter entered service with the Armée de l'Air in October 1978, the type being adopted as a short take-off and landing (STOL) design to fulfil a wide range of light and medium transport roles, including paradropping. It offered many operational advantages, including the same STOL capabilities as the much smaller MH-1521 Broussard then in service and nearly the same load-lifting capabilities as the Nord 262 that did not routinely land on unprepared strips. In all, ten Twin Otters were delivered to the French Air Force.

The Twin Otter has had a busy operational life with the Armée de l'Air, serving with a large number of units from bases in New Caledonia, Tahiti, Djibouti, and even Berlin, during the Cold War. For years, the Twin Otter has supported the United Nations Multinational Force and Observers (MFO) deployed to Sinai, a peacekeeping force, two aircraft being tragically lost in accidents there. From 2009 to 2012, a single Twin Otter was flown from Cayenne, in French Guiana, taking advantage of its STOL capabilities to fly to remote communities in the Amazon forest. It was eventually supplanted by an additional CN235.

Out of five still in operational service today, two are allocated to Escadron de Transport 3/61 'Poitou', the AAE special operations airlift squadron stationed in Base Aérienne 123 Orléans-Bricy. They carry out light transport missions in support of deployed special forces teams and take an active role in the paradropping training of special operators. The DHC-6 can also be used for medical evacuations: a dedicated medical module has been specially developed for the Twin Otter, allowing emergency surgery to be undertaken while in flight, thus considerably improving the chances of survival of a wounded service personnel.

Although it is never officially confirmed by the AAE, it is widely known that the secretive Groupe Aéromobile 56 'Vaucluse' (GAM 56) at Base Aérienne 105 Évreux-Fauville uses three Twin Otters for special transport missions.

In French service, the Twin Otter has proved to be extremely reliable, with a rugged airframe and strong engines. Its STOL capabilities are unmatched (except by the smaller French Army Aviation Pilatus PC-6), the Canadian twin being able to operate in incredibly tight spaces or from rough dirt strips to deliver and recover troops and equipment. The type is known to have been fitted with optional armour kits to provide the pilots with some level of protection against small arms. The DHC-6 is well appreciated by pilots who like 'real aircraft' with conventional controls.

A Twin Otter photographed between two training sorties.

The Twin Otter is regularly used for parachute training, as depicted here.

The Twin Otter is equipped with a rather unusual yoke.

Casa CN235

With 27 aircraft firmly in service with the Armée de l'Air et de l'Espace, the Casa CN235-200/-300 has gradually taken a leading role in the utility, tactical airlift, medical evacuation and paradropping missions, carrying up to 44 paratroopers. It performs a large range of transport missions in continental France, in French overseas territories and further afield. The CN235 is filling a gap between the TBM700 on one side and the C-130 and the A400M on the other.

The French Air Force initially purchased a batch of 20 CN235 airlifters to supplement ageing C160 Transalls and C-130H Hercules, with service entry in 1993. The type also progressively supplanted, in the liaison and utility roles, the Nord 262s, which were eventually withdrawn from Armée de l'Air service in 2004. The CN235 is capable of carrying a 5,000kg (11,013lbs) payload over 900km (486 nautical miles), or 3,000kg (6,608lbs) at 2,500km (1350 nautical miles). Its cargo hold is large enough to accommodate an M53 turbofan, the engine that powers the Mirage 2000 fighter. Tragically, one CN235 was lost in a fatal crash in the Pyrenees in 2003.

The delays impacting the A400M programme led to the decision to procure an additional batch of eight CN235-300 airlifters to help maintain French airlift capabilities at a time when the withdrawal of the older Transalls had already begun. The new variant offers improved performance thanks to a beefed-up airframe, more powerful engines, modified propellers and improved pressurisation: in a hot and high environment, the new aircraft retains more power and can operate at higher weights, with a heavier payload. Moreover, it is fitted with a considerably more modern glass cockpit relying on EFIS instrumentation, with two recent flight management systems offering GNSS capability, and with an EGPWS (Enhanced Ground Proximity Warning System) with 3D database. It can be easily identified thanks to the new twin-wheel forward landing gear, the earlier CN235-200 being fitted with a single nose-wheel. The new landing gear offers better capabilities to operate from austere strips. Finally, the -300 was equipped from the outset with armour plating to protect the crew (-200 aircraft have now been back-fitted with the armour).

The two squadrons in Base Aérienne 105 Évreux, Escadron de Transport 3/62 'Ventoux' and ET 1/62 'Vercors', both operate the two versions, but pilots have to be qualified on each variant as they are significantly different from each other. One aircraft is permanently fitted with 32 airline-type seats for VIP transportation, and the rest of the fleet is engaged in logistics and training missions in France and abroad.

The Casa plays a crucial role in French overseas territories, including for SAR missions (the type has been cleared to drop rescue equipment and life-rafts from its rear ramp, a major advantage for SAR operations in remote areas). Ten are currently forward deployed in French overseas permanent bases:

Two CN350-200s in DA190 Tahiti, in the Pacific, in French Polynesia, with ET 82 'Maine' since 1996.

Two CN350-200s in BA186 Nouméa, in New Caledonia, also in the Pacific. They have served with Escadron de Transport 52 'Tontouta' since 1998.

Three CN350-200s in BA 367 Cayenne, French Guiana, with ET 68 'Antilles-Guyane' since 2012.

Escadron de Transport 88 'Larzac', in Djibouti, with one aircraft drawn from one of the two squadrons based in continental France.

Two CN235-300s with Détachement Air 181 La Réunion, in the Indian Ocean, with ET 50 'La Réunion'.

Series -300 aircraft have been chosen for La Réunion to replace Transalls because they offer better range and airfield performance than series -200s, allowing them to resupply remote outposts on the Iles Éparses (literally 'scattered islands'), a group of five isolated French islands – Europa, Juan de Nova, Bassa de India, Glorioso and Tromelin – spread across the Indian Ocean, around Madagascar.

French Army paratroopers stand ready to jump out of a CN235 during a training exercise.

In French service, the CN235 is often called the 'Transallito' in reference to the larger Transall that was withdrawn from use in 2022.

The CN235-200 did not have the range for the mission and the A400M would have been too powerful, but with the advent of the CN235-300, the Armée de l'Air has found the ideal tool to fill the gap in the region.

The Casa 235 has now been operating in the combat support / logistics role in Africa for a while: the first deployment took place in 1994 when France intervened in Rwanda. With combat operations taking a new dimension in Mali in 2013, the CN235 was drafted in for intra-theatre airlift and aeromedical evacuation. The type has been cleared to drop heavy pallets, up to 800kg (1,762lbs) for resupply missions to isolated patrols deep in the desert. Since 2013, the Casa has conducted medical evacuations of wounded or critically ill patients. Called 'Casa Nurse' when configured for Medevac, it has proved ideal for the role as it can operate from semi-prepared runways, including dirt and compacted sand strips in the desert.

This Casa CN235 is photographed at Hyères naval air station, in October 2012.

The CN235 has proved invaluable for a number of missions, including logistics' transport, medical evacuation and paratroopers training.

This CN235 takes off from Salon-de-Provence, in May 2022. The front wheels have fully retracted but the main gear is still travelling.

This Casa CN235-300 manoeuvres on the ground at Salon-de-Provence in May 2022.

Dassault Aviation Falcon 900

The Armée de l'Air et de l'Espace has long been a strong supporter of Dassault Aviation, operating many generations of fighters and transport aircraft. The French Air Force has successively operated MD311/312/315 Flamants and Mystère 20 business jets in a wide variety of roles. With the entry into service of the three-engine Falcon 50 in 1979, the Armée de l'Air had obtained an aircraft that offered suitable performance in the executive transport role. The four Falcon 50s were eventually transferred to the Marine Nationale and modified into maritime surveillance aircraft.

Two publicity-shy Falcon 900 trijets entered service with the Armée de l'Air in November 1987. The Falcon 900 can carry up to 13 passengers over 6,500km (4,039 miles) in very comfortable conditions, as expected for a business jet. It can also carry out medical evacuation missions, with aeromedical specialists provided by Escadrille Aérosanitaire 6/560 'Etampes', also stationed in Villacoublay. While the earlier Falcon 50 could only carry one wounded or critically ill person, the bigger Falcon 900 can carry two at a time. The switch from a VIP configuration with executive seating to a medical configuration takes only a couple of hours. One of the Falcon 900s or 2000s is held at three-hour readiness in case it should be needed for a medical evacuation anywhere in the world.

The Falcon 900s have been successively flown by Groupement des Liaisons Aériennes Ministérielles (GLAM), Escadron de Transport, d'Entraînement et de Calibration 65 (ETEC 65) and Escadron de Transport 60 (ET 60) from Base Aérienne 107 Villacoublay, near Paris. They can be easily visually identified by counting the 12 main cabin windows.

This anonymous looking Falcon 900 is seen overflying central Paris during a Bastille Day fly-past rehearsal. Noteworthy is the Eiffel Tower in the background. (Photo Olivier Ravenel, French Air and Space Force)

This shot shows the beautiful lines and the smart colour scheme of the Falcon 900. (Photo Jean-Luc Brunet, French Air and Space Force)

Dassault Aviation Falcon 2000

The Armée de l'Air has always been extremely satisfied with its Mystère 20s, its Falcon 50s and its Falcon 900s, all of which proved reliable and dependable, so purchasing the Falcon 2000 as the next VIP aircraft was an obvious choice and an easy step. Like the earlier Mystère 20, the Falcon 2000 is a twin, but a much bigger and longer-range one than its predecessor, with advanced aerodynamic features to guarantee short-field performance while still offering a very good range.

Like the two Falcon 900s, the two Falcon 2000s play a crucial role conducting long-range medical evacuations of critically ill or seriously wounded military personnel. Falcon 900s and 2000s take

The Falcon 2000 is a smaller aircraft than the Falcon 900 or Falcon 7X. (Photo Jean-Luc Brunet, French Air and Space Force)

While the Falcon 900 and 7X are trijets, the Falcon 2000 is a twin. (Photo Jean-Luc Brunet, French Air and Space Force)

turns handling the medical evacuation readiness. They also regularly appear on news bulletins when bringing home French citizens who have been taken hostage abroad by terror groups.

Alongside the Falcon 900s and 7Xs, they are flown by Escadron de Transport 60 at Base Aérienne 107 Villacoublay. It is worth noting that a maritime surveillance variant of the Falcon 2000, to be named Albatros, will enter service with the French Navy from 2025 to replace the five Gardian and eight Falcon 50 maritime surveillance aircraft.

Dassault Aviation Falcon 7X

Two Falcon 7X business jets are in service with the AAE for high-end VIP missions. They are mainly used to transport the French President or the French Prime Minister during official trips in France and abroad. They have replaced two Airbus A319CJ (corporate jets) that were operated by the Armée de l'Air from 2002 to 2010 before being sold to the civilian market. The two Falcon 7Xs were delivered in July 2009 and May 2010 respectively. While they are smaller than the A319CJs, they are less costly to operate while still offering impressive performance, especially in terms of range.

Like the Falcon 900, the 7X is a tri-engine jet that can easily operate from short runways. It was developed by Dassault Aviation as a high-end aircraft to compete with the products of Gulfstream and Bombardier. Although it looks quite similar to its predecessor, it is, in fact, much bigger and offers a much longer range, a higher payload and an even more comfortable main cabin for the passengers, with a considerable lower cabin altitude to reduce aircrew and passenger tiredness during very long distance flights. The aircraft is equipped with a modern range of communication systems that allow the passengers to remain in contact with French authorities anytime, anywhere. It was the first business jet in the world to be fitted with fly-by-wire controls and side-sticks.

Like their Falcon 900 and Falcon 2000 brethren, the Falcon 7Xs are operated by Escadron de Transport 60 from Base Aérienne 107 Villacoublay, west of Paris. While the A330AUG is now the main presidential aircraft for long-distance trips, the Falcon 7Xs are often drafted in to carry additional personnel or media members on official trips, or to be used as reserve should a technical problem occur with the A330. The Falcon 7Xs regularly become the main presidential choice for shorter trips within Europe, or for flights to airports or airfields with shorter runways. They can easily be identified by counting the 14 main's cabin windows.

To distinguish a Falcon 900 from a Falcon 7X, just count the windows: 12 on the former and 14 on the latter. (Photo Jean-Luc Brunet, French Air and Space Force)

This Falcon 7X is photographed at Hyères Naval Air Station, close to the Fort de Brégançon, the French Président's summer residence.

The Falcon 7X has proved highly successful as an official aircraft, having been selected by a number of nations for VIP transportation. (Photo Richard Nicolas-Nelson, French Air and Space Force)

Socata TBM700

The Socata (now Daher) TBM700 is used by the AAE as a fast liaison and calibration aircraft. The TBM700 can carry up to four passengers in comfortable conditions anywhere in France and further afield at affordable cost and within a very short timeframe. The type was initially procured to replace the Morane-Saulnier MS760 Paris, a two-engine, four-place jet that guzzled fuel at an alarming rate for such a small aircraft. The TBM is nearly as fast as its predecessor, and can carry six people while burning considerably less kerosene thanks to a single, fuel-efficient Pratt&Whitney Canada PT6A-64 turboshaft engine rated at 700 shp. Initially, some Armée de l'Air personnel were concerned that a single-engined aircraft would not be safe enough to operate in the VIP transportation role, but they were proved wrong thanks to the legendary reliability of the PT6 turbine: the TBM family (from the TBM700 and all its successors right down to the current TBM960) have an impeccable safety record and no more than a couple of aircraft have been lost worldwide even though more than 1050 examples had been delivered at the time of writing.

While the MS760 Paris only had a basic avionics suite, the TBM700 has been fitted from the outset with modern systems, including an advanced autopilot and a weather radar. It can theoretically be flown by a single pilot, but this is not often the case within the AAE. It is capable of operating day and night in foul weather, with only 150 metres (492 feet) of horizontal visibility and 61 metres (200 feet) of ceiling. It is renowned for its short runway take-off and landing capabilities. It has enough range to fly direct from Paris to Athens, for example, at an average cruise speed of 280 knots while burning very little fuel. Equally importantly, passengers are accommodated in good conditions, the aircraft proving extremely comfortable even for maximum range flights. Its only real drawback is that it is not fitted with toilets, sometimes a problem for the longer-range missions across Europe.

The type entered French Air Force service in 1991 and, today, all 15 TBM700s remain in service with the AAE. Eight are serving with Escadron de Transport 41 'Verdun', at Base Aérienne 107 Villacoublay, with a further six with Escadron de Transport 43 'Médoc', at Base Aérienne 106 Bordeaux-Mérignac. The final aircraft is allocated to the Centre d'Expertise Aérienne Militaire (CEAM, military air expertise centre, also known as the French air warfare centre), at Base Aérienne 118 Mont-de-Marsan, where it conducts a wide range of transport and trial missions in direct support of the operational evaluation work performed by the CEAM.

The TBM700 fleet has proved extremely useful for a wide range of liaison and light transport missions.

Above left: French generals and decision makers often rely on the TBM700 to be on time for important meetings at the other end of the country.

Above right: In French Army and Air Force service, the TBM700 has proved to be a reliable and dependable aircraft.

Right: An anonymous TBM700 taxies out at Orange, in June 2022.

Below: This TBM700 is photographed at Salon-de-Provence between two sorties. The cockpit is protected from the scorching heat by a dedicated cover.

Chapter 4
Helicopters

Sud Aviation SA330 Puma

The SA330 Puma has long been the workhorse of the French Army Aviation and of the French Air Force helicopter component. The rotorcraft was jointly developed with the UK as part of a broader industrial, military and political co-operation that also included the nimble Gazelle and the successful Lynx (not forgetting the beloved Jaguar). A total of 37 SA330 Pumas were eventually purchased by the Armée de l'Air, with service entry in May 1974. Over the years, the type has carried out a large range of missions, including tactical transport, presidential transport, search and rescue (SAR), special forces support, counter terrorism, and combat search and rescue (CSAR). The Puma may even be used for paradropping missions. Armée de l'Air Pumas have successfully served in a wide range of environments, from the damp jungle of French Guiana to the sandy and dusty desert of Djibouti and from salty French islands in the Pacific to the high, wind-swept mountains of continental France and Corsica, proving to be rugged and dependable helicopters in those demanding conditions.

In the 1990s, a number of Pumas were modified for CSAR operations with armour plating, a Chlio FLIR turret, a door-mounted machine-gun (only one, on one side), an NVG-compatible cockpit, a beacon locator and a comprehensive self-protection suite composed of a Damien A missile approach warner and of Éclair or Saphir decoy dispenser. They were also fitted with large undercarriage sponsons holding additional fuel tanks, which brought overall fuel capacity to 2,250 litres (495 gallons) instead of 1,550 litres (341 gallons). In early 2004, the CSAR Pumas were armed with the MAG 58 7.62-mm machine gun that supplanted the outdated ANF1. These highly-specialised Pumas were in service with Escadron d'Hélicoptères 1/67 'Pyrénées', based at Cazaux, the unit that had become the dedicated CSAR squadron in the 1990s.

Although still a sprightly performer, the Puma is now fast ageing. The type was withdrawn by Escadron d'Hélicoptères 1/67 'Pyrénées' in 2017 when the squadron became an all-Caracal unit, but it remains in service with a dwindling number of squadrons, mainly overseas.

In Base Aérienne 126 Solenzara, in Corsica, Escadron d'Hélicoptères 1/44 'Solenzara' is the island's dedicated search and rescue crew. It can also perform tactical missions and it trains all future Puma aircrews.

In French Guiana, Pumas allocated to Escadron de Transport 068 'Antilles-Guyane' take an active role fighting illegal gold exploitation by 'garimpeiros' gang members from nearby Brazil. Illegal gold mining is a major problem in the vast French Amazonian territory (which is larger than Scotland), creating chronic insecurity and growing pollution due to the massive and uncontrolled usage of mercury, fuel and heavy metals in the prohibited mining process. Pumas are tasked with carrying groups of gendarmes and commandos whose task it is to destroy the mining and support infrastructure used by the 'garimpeiros'.

In Djibouti, Escadron de Transport 88 'Larzac' provides search-and-rescue coverage and tactical transport to French units based in this small east African country, often operating in difficult conditions in a harsh desert environment. AAE pilots posted to Djibouti soon become experts in landings in brown-out conditions.

In New Caledonia, Escadron de Transport 52 'Tontouta' operates three Pumas for SAR and transport missions, supporting the civilian population and French military forces in this archipelago in the southern Pacific. Stationed in Base Aérienne 186 Nouméa, the unit is a joint Puma / CN235 squadron.

Finally, in Base Aérienne 107 Villacoublay, near Paris, the Groupement interarmées d'hélicoptères (GIH, or joint helicopter group) supports the GIGN (Groupe d'Intervention de la Gendarmerie Nationale) and the RAID (Recherche, Assistance, Intervention, Dissuasion, or search, help, intervention, deterrence) counter-terrorism units. The joint GIH operates a mixed fleet of Army and Air Force Pumas. Its aircrews are cleared to use special tactics for the insertion of operators and the recovery of hostages and gendarmes and policemen, relying on Special Patrol Insertion/Extraction (SPIE) rope and dedicated platforms carried as sling-loads.

Right: This GIH Puma is photographed with a German Bundespolizei Super Puma during a counter-terrorism exercise in Germany, in 2014.

Below: The GIH occasionally cooperates with foreign units for counter-terrorism operations as part of the European Atlas network of units. Here, a Puma is flying in close formation with a German Bundespolizei Super Puma.

A dark grey-painted CSAR Puma taxies in at Tours during an airshow in 2009.

This Escadron d'Hélicoptères 1/44 'Solenzara' Puma comes in to land after a training mission, in 2016.

This head-on shot highlights the enlarged fuel sponsons that carry additional fuel for the CSAR variant.

Aérospatiale AS332 Super Puma

After the successful introduction of the Puma, it was logical for the Armée de l'Air to adopt the next evolution of the aircraft, the Super Puma / Cougar. The first three short-fuselage AS332C Super Pumas were delivered to the Direction des Centres d'Expérimentations Nucléaires (DirCEN, or nuclear experiment centres directorate) from 1984 to support French nuclear trials in the Pacific. They were flown and maintained by Air Force personnel and were eventually joined by a fourth aircraft purchased in 1989.

Two long-fuselage AS332L Super Pumas were acquired in 1988 to replace the Pumas then used for the VIP presidential transport role within the Groupement des Liaisons Aériennes Ministérielles (GLAM), the then dedicated presidential and ministerial transport group stationed in Base Aérienne 107 Villacoublay, near Paris. A third one was added later to the inventory. They all transferred to Escadron d'Hélicoptères 3/67 'Parisis' (still in Villacoublay) a few years later.

A further batch of three stretched Super Pumas was acquired to equip the Groupe Aéromobile 56 'Vaucluse' (GAM 56) based in Base Aérienne 105 Évreux-Fauville to support clandestine operations. These aircraft were called Cougars by the Armée de l'Air. Their missions and equipment fit were classified.

Once the French nuclear test centre in the Pacific shut down, the DirCEN Super Pumas remained in service in the Pacific for SAR missions. They later transferred to continental France to supplement the older Puma in the SAR role, the rescue missions in the Pacific being taken over by the French Navy. All rescue and special missions' Super Pumas / Cougars have now been withdrawn from service and sold on the second-hand market (including two to the Spanish Air Force), the Cougar only remaining in service with the Aviation Légère de l'Armée de Terre, the French Army Aviation.

Three Super Pumas remain in AAE service for VIP transport missions from Villacoublay air base. They are equipped with an air-conditioning pack and with an executive cabin, with comfortable seating and an extensive communication suite. More importantly, they are fitted with a comprehensive avionics suite and with de-icing equipment that allows them to fly in (nearly) all weathers, a crucial advantage when transporting the French President, the French Prime Minister or foreign heads of states on official trips in France. The three VIP Super Pumas are now allocated to Escadron de Transport 60, the dedicated unit for ministerial transportation role. AAE pilots flying these Super Pumas do not wear regular flight suits during official trips, and instead fly in shirtsleeves, like civilian airline pilots.

One of the three long-fuselage AS332L Super Pumas used for VIP transportation lifts off from a football pitch in a small village after an official ceremony.

This AS332L Super Puma is photographed at the Paris Air Show where it has taken French President Emmanuel Macron for an official visit.

The massive system on the side of this VIP transportation AS332L Super Puma is an air-conditioning pack, a welcome addition for the comfort of the dignitaries routinely transported by these rotorcraft.

Aérospatiale AS555AN Fennec

The Fennec is the military variant of the acclaimed civilian Ecureuil / Squirrel range. The French Air Force purchased eight AS355F1 Ecureuils and 43 AS555AN Fennecs that eventually supplanted all the Alouette II and III helicopters that had previously been operated since the 1950s. The eight Ecureuils were withdrawn early, falling victim to budget cuts but, out of 43 Fennecs delivered to the Armée de l'Air from 1990, 40 remained in service at the time of printing, forming a very effective multirole rotary force capable of performing a wide range of roles, in continental France, in French overseas territories and abroad. They can be fitted with a winch for SAR missions. Fennecs have been deployed to Libreville, in Gabon, and to Abidjan, in Ivory Coast, for more than 30 years, providing support to French forces permanently based in these two African countries.

The French Air Force is unique in that it maintains five AS555AN Fennec helicopters ready to react from four of its air bases: Saint-Dizier, Orange, Bordeaux and Villacoublay (two). These locations have been carefully chosen to provide air-defence against slow movers to strategic cities and targets such as Paris, Bordeaux, various nuclear power plants and other key industrial facilities. Like the fighters, the Fennecs are maintained at short readiness, day and night. For the air-defence mission, each Fennec is flown by a crew of four: pilot, co-pilot and two snipers, one of them qualified as team leader and the other as a shooter. The snipers are trained to very high standards. They carry two types of weapons: a HK417 7.62-mm semi-automatic precision rifle (that has now totally replaced the bolt-operated FRG2 rifle) and a 12-gauge FN Herstal TPS pump-action shot gun, which can fire either red or green flares, or more lethal nine-pellet buckshot ammunition, ideal to engage a micro-drone, a paramotor or a paraglider. With the 7.62mm rifle, snipers claim a precision of 50cm (20in) when firing at a target 200 metres (650 feet) away while flying at 110 knots. For missions at night, French Air Force Fennecs

The nimble Fennec performs unglamorous but much needed air-defence tasks, standing ready to intercept unknown slow movers overflying forbidden areas.

can be equipped with either a Chlio or an Ultra 7000 FLIR turret in addition to the crew's night vision goggles, allowing intruders to be identified at stand-off distances. The ageing Chlio and Ultra 7000 pylon-mounted turrets are due to be replaced from late 2023 by a nose-mounted Trakka TC-300 FLIR that will offer considerably better performance and less drag. The Fennec is also cleared to carry an axial-firing M621 20mm cannon with 240 shells in ammunition boxes in the main cabin and a dedicated gun sight in the cockpit. This gun is seldom carried, however, except in French Guiana where Fennecs are used to protect the Ariane rocket launches from the Kourou space-port. Finally, Fennec-borne snipers are also trained to provide air-to-surface precision / suppressing fires. They can hit their targets, including fast moving objects such as cars, with utmost precision.

The Fennec is operated by Escadron de Transport 43 'Médoc', at Base Aérienne 106 Bordeaux-Mérignac, Escadron d'Hélicoptères 3/67 'Parisis', at Base Aérienne 107 Villacoublay, Escadron d'Hélicoptères 1/65 'Alpilles' and the Centre d'Instruction des Équipages d'Hélicoptères (helicopter crew training centre) 00.341 'Colonel Santini', both at Base Aérienne 115 Orange-Caritat, and by Escadron de Transport 068 'Antilles-Guyane', at Base Aérienne 367 Cayenne Rochambeau, in French Guiana.

Although not the best known AAE aircraft, the Fennec plays a crucial role behind the scenes. Under current plans, the Fennec fleet will be replaced by the new H160M Guépard that will offer higher speed, longer range, increased payload and a contemporary avionics suite. The AAE has expressed a need for 40 Guépards in addition to another 80 for the French Army Aviation, 49 for the French Navy and one to be used as a flying test bench by the DGA, the French defence procurement agency.

Even though the Fennec can be used for SAR, the rescue winch is seldom carried by the Fennecs in continental France.

With a marksman at the door for a training session, this Fennec manoeuvres hard over the Diane firing range, in Corsica.

This Escadron d'Hélicoptères 5/67 'Alpilles' Fennec is flying is appalling weather close to its Orange base.

Both the French Air Force and the French Army Aviation field Fennecs for a wide range of operational and training tasks.

The Fennec fleet is spread over bases in continental France and in French overseas territories.

Eurocopter EC725 / H225M Caracal

The EC725 programme was launched by the Armée de l'Air in the late 1990s as an effort to find a new rotorcraft that would specialise in the Combat Search and Rescue (CSAR) mission and eventually replace the ageing Puma in that role. Compared to the Puma and the Super Puma, the new type would have to be more capable, with a longer range and with in-flight refuelling capabilities. The resulting EC725 is heavier and more powerful, giving a better payload thanks to more powerful engines and a five-blade main rotor.

The EC725 can easily be distinguished from the Cougar thanks to its new five-blade rotor and spheriflex rotor head, which offer a 20 per cent increase in lift and boost maximum admissible take-off weight from 9,700kg (21,365lbs) to 11,200kg (24,669lbs). Additionally, the EC725 is powered by improved Turboméca Makila 2A engines each rated at 1,566 kW (2,100shp) for take-off. The two turbines are each equipped with a dual channel FADEC (Full Authority Digital Engine Control) and, to absorb the extra power, they are linked to an uprated main gearbox with a 30-minute dry-run capability. For autonomous operations 'in the field', an auxiliary power unit is fitted. Even more importantly for the CSAR mission, it offers a slightly higher speed than its predecessors, and a considerably longer range. Its fuselage was lengthened, allowing the adoption of two additional windows at the front of the cabin that could be used to install window-mounted MAG58 7.62mm machine-guns for self-defence and fire-suppression to keep heads down while recovering an ejectee behind enemy lines (on both sides, a major improvement over the Puma whose single machine gun covered only one side). For missions requiring more fire power and a longer reach, the Caracal can now carry the SH20 20mm cannon in the main cabin, firing through the right main door aperture.

The Caracal was equipped from the start with a state-of-the-art avionics suite that included a nose-mounted Chlio ST FLIR turret and six multi-function displays, and with a comprehensive self-defence/electronic warfare suite for maximum survivability in a contested environment. The Chlio turret was already an antiquated system when the Caracal entered service and it was felt that it did not provide adequate performance levels. As a consequence, a programme was initiated by the Armée de l'Air, in 2015, to replace it with the more modern Euroflir 350, designed and produced by Sagem (now Safran). It comprises the latest generation sensors to expand the Caracal's operational capabilities at night and in marginal weather conditions.

The Caracal is the only French rotorcraft that can be refuelled in flight. The lack of dedicated tankers initially led the French Air Force to operate with Italian KC-130J Super Hercules and USAF MC-130 Hercules, but the recent entry into service of two KC-130Js in France has radically changed the situation, allowing the AAE to undertake long-range Caracal missions independently, although regular training is still conducted with the RAF Mildenhall-based MC-130J Commando II tankers of the US Air Force Special Operations Command. It is expected that the A400M Atlas will also be capable of refuelling Caracals in the not-too-distant future, once all technical hurdles have eventually been cleared.

AAE Caracals are all allocated to Escadron d'Hélicoptères 1/67 'Pyrénées' at Base Aérienne 120 Cazaux. The squadron belongs to the Brigade des Forces Spéciales Air (BFSA), the AAE's special forces brigade. As such, it supports special forces operations in France and abroad, but also stands ready to perform CSAR missions or conventional SAR missions. The CSAR role also includes deployments on board nuclear aircraft carrier *Charles de Gaulle*. For CSAR missions, AAE Caracals closely co-operate with the specialists of the Commando Parachutiste de l'Air No 30 (CPA30, or Air Force Parachute Commando No 30), the leading French commando unit for CSAR recoveries in France. All its commandos undergo a very long screening and training process before being qualified to undertake a live CSAR mission. They are all cleared to use the Special Patrol Insertion/Extraction (SPIE) rope for insertion into tight spaces and/or for recoveries from confined areas.

Helicopters

The first Armée de l'Air order included ten EC725s, one of which was destroyed in an accident in 2014. A replacement aircraft was ordered in 2019, with the delivery being expected in 2023, at the time of writing. A further eight, now called H225M by Airbus Helicopters, were purchased at the beginning of the Covid crisis as part of a series of measures to support the French economy, all of which should be in service by 2025. These new aircraft will allow some of the fast-ageing Pumas to be withdrawn from use while increasing the inventory of heavy lift, special ops / CSAR helicopters. The Caracal has a huge growth potential and other equipment and weapons could be added in the future, such as laser-guided 68mm rockets and AGM-114 Hellfire air-to-surface missiles.

An EC725 photographed off Cazaux. Its in-flight refuelling probe and its FLIR turret are clearly visible.

This Caracal has extended its in-flight refuelling probe during an airshow routine. The gunner is waving at the crowd.

This Caracal will soon be refuelled off Cazaux by a USAF MC-130J Commando II of the 67th Special Operations Squadron 'Night Owls' / 352nd Special Operations Wing from RAF Mildenhall, in Suffolk.

This EC75 Caracal is photographed between two sorties at its Cazaux home base. The Caracal is one of the most reliable rotorcraft in service in France.

This EC725 Caracal comes in to land during a CSAR tactical demo. It will shortly recover and evacuate to safety a fast jet pilot after a simulated ejection and recovery.

The Caracal's in-flight refuelling probe is a massive piece of kit. All Caracal pilots need to be qualified for in-flight refuelling.

Eurocopter EC225

The two EC225 helicopters in service with the Armée de l'Air et de l'Espace were initially purchased from Eurocopter by the French Navy as part of an emergency procurement programme. A need had appeared for an interim fleet to fill the gap between the three-engine SA321G Super Frelon and the then future NFH90 Caïman, a rotorcraft that had been severely delayed due to budget constraints after the fall of the Berlin Wall and the end of the Cold War. At the time, the fast-ageing Super Frelon that equipped Flottille 32F at Lanvéoc-Poulmic was experiencing availability and even flight safety issues (an electric fire had seriously damaged one aircraft, triggering questions about the reliability of the aged airframes). As a result, the decision was made to withdraw the iconic three-engine rotorcraft and replace it with two EC225s, a modern type that would offer good performance levels, advanced systems and excellent availability rates. The first EC225 was delivered to Flottille 32F in April 2010.

The EC225 (now marketed by Airbus Helicopters as the H225) proved highly successful with the French Navy, performing SAR missions from Lanvéoc-Poulmic and Cherbourg, often in appalling weather conditions. When the NH90 Caïman finally entered service with Flottille 33F at Lanvéoc-Poulmic, the two EC225s became surplus to requirement. They were withdrawn from Navy service and temporarily stored. Consequently, Flottille 32F disbanded (only to be recreated in 2022 as the first H160 operator in France).

The story was far from over for these two EC225s, as the secretive Groupe Aéromobile 56 'Vaucluse' (GAM 56) based in Base Aérienne 105 Évreux-Fauville expressed a requirement for the transfer of the two stored rotorcraft to replace older and less capable Super Pumas. That move

The two EC225s currently serving with the GAM 56 started their career with the French Navy as 'gap fillers' between the Super Frelon and the NH90 Caïman.

Above: For additional range, the two EC225s can carry external fuel, a crucial advantage for the kind of missions they now routinely undertake.

Right: Head-on view of the EC225 clearly showing the rescue winch that can be fitted to the right side of the airframe.

was eventually approved and they were reassigned from Marine Nationale to Armée de l'Air ownership in 2016. The EC225 is equipped with a winch and can be fitted with external fuel tanks for added range and endurance, a decisive advantage for the type of clandestine missions routinely undertaken by the GAM 56. Besides, it offers a very high level of commonality with the EC725 Caracal already in service with Escadron d'Hélicoptères 1/67 'Pyrénées' at Base Aérienne 120 Cazaux, thus considerably easing service entry, facilitating air and ground crew conversion training and allowing the pooling of some of the maintenance for maximum operational efficiency.

The EC225 is undoubtedly a sprightly performer. The French rotorcraft is surprisingly agile for such a large aircraft.

Chapter 5
Trainers

Dassault Aviation / Dornier Alpha Jet E

The Dassault Aviation / Dornier Alpha Jet is the result of a collaborative project between France and Germany. The programme had been launched in the 1970s to provide an advanced trainer to the French Air Force in order to replace French Fouga Magister, T-33 and Mystère IVA trainers, and a light attack aircraft to the Luftwaffe to supplant G.91 light strike fighters. The new design was fitted with a modern avionics suite (modern for the time), a tandem cockpit, and with two engines for enhanced safety. It could carry a hefty payload under four wing stations and could be armed with a single cannon in an external pod mounted under the fuselage (27mm BK27 gun on German aircraft, and 30mm DEFA cannon on French ones). Two separate variants were produced for the two countries: Alpha Jet E (École, meaning 'school') for the Armée de l'Air and Alpha Jet A (Attack) for the Luftwaffe. In all, 175 Alpha Jets entered French Air Force service from 1979. Initial feedback was excellent: the plane was remarkably agile and more representative of the new fighters then in service (Jaguar and Mirage F1), but some former T-33 and Mystère IVA instructors complained that the Alpha Jet was far too easy to fly and land!

While Germany withdrew its Alpha Jets after the end of the Cold War, France kept utilising the type for a wide range of training and aggressor missions, including weapons' training firing from Cazaux (with 68mm rockets, practice free-fall bombs and a 30mm cannon in the air-to-ground role and a 30mm gun in the air-to-air role). In the early 2010s, 20 Alpha Jets benefited from a limited upgrade for the training role, with a HUD (Head-Up Display), an inertial navigation system, and a multifunction screen. Three generations of French fighter pilots have successfully been trained on the Alpha Jet. In 2023, the AAE will stop using the Alpha Jet E as an advanced trainer, however, with that role being assumed by the more economical and more modern Pilatus PC-21.

The Alpha Jet has also long been used as an aggressor asset, providing efficient training at reduced costs compared with front-line fighters. Over its service life, various dedicated flights and squadrons have operated it in that role from Bases Aériennes Saint-Dizier and Dijon. Today, Escadron d'Entraînement 3/8 'Côte d'Or' flies the type in the aggressor mission from Base Aérienne 120 Cazaux, providing sparing partners to front-line Rafale and Mirage 2000 squadrons and allowing basic fighter manoeuvring training to be advantageously conducted. The nimble Alpha Jet is renowned for its agility and when flown by an experienced pilot it becomes a potent adversary at close range, even for a Mirage 2000 or a Rafale. Its operating cost is about half that of the Mirage 2000, and about one third that of the Rafale. At low speed, its sustained turn rate is better than that of a Mirage 2000, a decisive advantage for the quality of the training. The Rafale is more manoeuvrable than the Alpha Jet, however. Being fully capable of performing close air support training, the Alpha Jet has proved tremendously useful in the forward air controller training role. It is often flown with two drop tanks, giving 2 hours 30 minutes endurance. With the drop tanks, the aircraft is limited to 5 g.

Stationed in Base Aérienne 701 Salon-de-Provence, the Patrouille de France is probably the best known Alpha Jet operator anywhere in the world. This prestigious unit has been equipped with the type since 1981 when it supplanted the classic Fouga Magister in service with the team since 1964. Apart from its tri-colour paint scheme, Patrouille de France Alpha Jets differ from their counterparts with the adoption of a smoke system and its control panel in the cockpit; a powerful light in the nose to facilitate visual acquisition during head-on passes by the solos; and the removal of the gunsight in the front cockpit to provide a better field of view in the forward hemisphere. Some of the aircraft are also fitted with an angle-of-attack (AOA) indicator on the left side of the cockpit instead of the right side, allowing their pilots (those flying on the left-hand side of the team) to know the AOA they are flying at without having to look at an instrument on the other side of their cabin. This is a modification that increases flight safety: the pilots flying on the left of the formation can briefly look at the AOA indicator as and when needed while still keeping their wingmen in sight. Although plans could still change, it is anticipated that the Patrouille de France will operate the Alpha Jet until 2035 at least. No replacement, or replacement programme had been announced at the time of writing. The AAE is carefully managing the Alpha Jet's remaining airframe life to ensure that the aircraft will reach its expected out-of-service date without encountering major technical hurdles.

Right: **The Alpha Jet is a remarkably safe aircraft. It also proves straightforward to maintain and repair.**

Below: **Photographed in 2012, this Cazaux-based Alpha Jet E belonged to the now-disbanded Franco-Belgium Advanced Jet Training School (AJeTS).**

The Alpha Jet E has also proved extremely useful as an aggressor platform, helping Mirage 2000 and Rafale pilots train at reduced costs.

Above: The Patrouille de France has operated the Alpha Jet for more than 40 years.

Right: Escadron d'Entraînement 3/8 'Côte d'Or' flies the Alpha Jet E in the aggressor role from Cazaux. This 'Côte d'Or' aircraft is fitted with two underwing drop tanks.

Below: Photographed at their Salon-de-Provence home base, these Patrouille de France Alpha Jets are neatly parked in front of their hangars.

Above: The Larzac engines have been started. The eight Patrouille de France Alpha Jets will soon taxy for another display.

Left: Patrouille de France Alpha Jets are numbered from 0 to 9, numbers 0 and 9 being the dedicated spare aircraft.

This Patrouille de France Alpha Jet taxies past the Rafale Solo Display Team's specially-painted Rafale C at Salon-de-Provence, in May 2022.

The Patrouille de France is the best known operator of the Alpha Jet E anywhere. The display team will continue with the type for the foreseeable future.

Pilatus PC-21

The propeller-driven Pilatus PC-21 has been adopted by the French Air Force as its next generation trainer for prospective fast jet aircrews. The selection of the Swiss design was controversial: as a new generation advanced trainer, it was perceived by many as a huge step backwards in terms of capabilities compared to the Alpha Jet, but the introduction of the type is now seen as a major step forward because of its state-of-the-art avionics suite with embedded mission simulation, its excellent (some say good enough) performance level and, even more importantly, its significantly reduced operating costs compared to the twin-engine Alpha Jet. PC-21 deliveries to the Armée de l'Air started in 2018 as part of a private finance initiative project, with all aircraft to be stationed in Base Aérienne 709 Cognac-Châteaubernard. The PC-21 has now replaced both the last few remaining TB30 Epsilons based in Cognac, and the Alpha Jet E trainers used for fighter pilot training in Tours Saint-Symphorien until 2020. At the time of publishing, its role was being expanded further to include tactics training, progressively taking over the missions of the Alpha Jets of the fighter weapons school at Cazaux.

The PC-21 is a state-of-the-art aircraft, made of modern materials and with a modern avionics suite. In comparison to the Epsilon and the Alpha Jet, it offers an advanced cockpit environment and the latest generation of systems, including an inertial navigation system (INS). Its glass cockpit comprises three large multifunction displays, a HUD (Head-Up Display) and HOTAS (Hands On Throttle And Stick) technology, with a layout representative of the latest generation of fighters. Moreover, the PC-21 is equipped with an OBOGS (on-board oxygen generation system), an anti-g system and with two Mk 16 zero-zero ejection-seats, a major advantage for aircrew safety over the Epsilon that was not fitted with ejection-seats. Although the type is not as fast as the Alpha Jet, it nevertheless offers impressive capabilities: it can cruise at 330 knots while still maintaining a very small fuel flow courtesy of its powerful and fuel-efficient Pratt & Whitney PT6 turbo-shaft engine rated at 1600hp (top speed is 370 knots). Thanks to hydraulically-assisted ailerons, roll spoilers and powerful elevators, the PC-21 is remarkably agile and can manoeuvre at up to 8 g. A digital power management system and automatic yaw compensation make the PC-21 easy to fly in the circuit, providing jet-like behaviour and feel. Its only real drawback is its rather high noise level that has attracted critics from the local population around Cognac. The French Air Force has accrued enough experience on the PC-21 to confirm that a propeller-driven aircraft fitted with the latest generation of avionics with embedded simulation can advantageously replace a fast jet trainer as a fighter lead-in trainer. Young pilots now get their wings after flying the PC-21 and go straight on to the Mirage 2000 or the Rafale without ever flying a jet before, in a syllabus very similar to that developed by the Swiss Air Force whose young pilots convert on to the F/A-18C/D Hornet right after graduating on the PC-21.

To reinforce dedicated air-defence operations, the PC-21 can also be engaged in the homeland defence mission, providing additional assets for the protection of major events such as the G20 summit or the D-Day anniversary celebrations: its pilots take advantage of its wide range of operating speeds to quickly intercept slow movers and maintain close formation with them in order to identify any unknown intruder.

The AAE is so happy with its PC-21s that it announced the procurement of a second batch of nine in July 2021, bringing the number on order to 26, which should all be in service by late 2023. By then, three squadrons of the École de l'Aviation de chasse 00.315 will be equipped with the Swiss trainer: Escadrons d'Instruction en Vol 1/13 'Artois', 3/4 'Limousin' and 3/13 'Auvergne'. All these units have been allocated the number plates and names of previously disbanded fighter squadrons.

A pair of Pilatus PC-21 trainers recovers into Cognac at the end of a training sortie. French PC-21s adorn a smart paint scheme.

This Pilatus PC-21 taxies out at Cognac for a training sortie. The PC-21 is an extremely modern and powerful trainer.

These two PC-21s are parked in front of a hardened aircraft shelter at Orange air base, in October 2021.

In the PC-21, the instructor in the back has excellent forward visibility.

The PC-21's carefully shaped wing tips can clearly be seen on this photo of a PC-21 taxiing out at Évreux in late 2021.

The PC-21's side-opening canopy proves to be a major advantage when ejection-seat maintenance is required. The seats can be removed and re-installed without taking the canopy away.

Embraer EMB-121 Xingu

The Embraer EMB-121 Xingu is the AAE's standard multi-engine training aircraft. It was purchased from Brazil as part of a bilateral agreement between the two countries. From May 1982, the Xingu began replacing the piston-powered Dassault Aviation MD312 Flamant twins that had been in Armée de l'Air service in the training role since the 1950s. It eventually also supplanted the Piper PA-31 Navajo then in service with the French Navy for various trials and support, training and liaison missions. The classic Flamant was by then already outdated and a replacement was urgently needed. The Navajo was a much smaller and far less capable aircraft. In all, 41 Xingus were purchased as part of a joint procurement between the Armée de l'Air and the Marine Nationale to equip both the French Air Force (25) and the French Naval Aviation (16) for a variety of training and transport missions.

When it entered service, the Xingu was unquestionably one of the most advanced multi-engine aircraft in service in the French military, far more modern than the Noratlas, Caravelle, Nord 262, DC-8 and P2V-7 Neptune still in widespread use in France. All these types had been introduced in the 1950s and 1960s and were fast ageing. In sharp contrast, the Xingu had the latest avionics at the time and was powered by modern and remarkably reliable PT6A-28 turbines, each rated at 1,360hp. It turned out to be ideal to train future Air Force Transall airlifters and Navy Atlantic 1 maritime patrol aircraft pilots.

Today, the Xingu remains in service with the École de l'Aviation de Transport 00.319 (EAT 319, or aviation transport school) 'Capitaine Jean Dartigues' at Base Aérienne 702 Avord for day and night tactical transport training. EAT 319 is the joint multi-engine school where all future French transport, in-flight refuelling, AWACS and maritime patrol/maritime surveillance pilots are trained to fly 'heavies'. No fewer than 22 Xingus are presently allocated to the joint school, some of them drawn from the French Navy. The aircraft, with its relatively short fuselage, is said to suffer from a longitudinal stability issue, especially at low speed in the airfield circuit. This problem has proved to be an advantage in the training role, however, the students having to fly a demanding aircraft at a crucial time in their career, thus helping develop advanced piloting skills, particularly in IFR conditions. Xingu pilots also have to be careful not to enter a manoeuvre that would cause the angle of attack to increase at the risk of disturbing the airflow around the elevators of the plane's T-tail. The Xingu is regularly displayed at airshows as part of the Camomille four-ship team usually flown with a mix of aircraft in Navy and AAE markings. The Xingu is also in service with Flottille 28F of the French Naval Aviation at BAN Lann-Bihoué, in south Brittany.

The Xingu fleet has benefited from a modernisation of its avionics suite in the mid-2010s to ensure that it remains compliant with the latest civilian regulations in order to keep on operating in all airspaces without any restriction for the foreseeable future. To reduce operating costs, Xingu maintenance is now undertaken by private contractors from Airbus CATS (Cassidian Aviation Training Services).

The Xingu is routinely tasked to perform light transport missions, taking advantage of the cabin that often remains empty during training missions. In 2020, at the peak of the Covid crisis, Xingus were drafted in to move medical teams from area spared by the illness to regions that were badly hit. The type celebrated its 40th anniversary in French service in 2022 and it undoubtedly remains a crucial asset in the French inventory, playing an unglamorous but much needed backstage role.

Strangely enough, the Xingu does not look outdated at all although it celebrated its 40th anniversary within the French Air Force in early 2022.

With its short fuselage, the Embraer Xingu has sometimes proved to be a tricky aircraft to fly well by student pilots.

At some stage, some Xingus were painted in the same overall grey paint used for a range of French tactical aircraft.

Grob 120A-F

The Grob 120A-F is used as a basic trainer by the AAE at Base Aérienne 709 Cognac-Châteaubernard, in the south-west of France. The German-made aircraft and the Pilatus PC-21 have replaced there the Socata TB30 Epsilon that had successfully been used to train three generations of Armée de l'Air pilots. The Grob 120A-F and the Epsilon flew alongside from Cognac for a couple of years, but that era finally came to an end in 2019 when the faithful TB30 was withdrawn from use after 35 years of sterling service, leaving the 18 Grob 120A-F trainers to share the ramp with the more-capable PC-21s. The Grob 120A-Fs are owned and maintained by Airbus CATS (Cassidian Aviation Training Services) but flown by military instructors as part of a private finance initiative project.

The Grob 120 is a derivative of the earlier Grob 115 adopted by a number of air arms, including the Royal Air Force (as the Tutor T1). Like the Epsilon, it is powered by a flat six Lycoming engine rated at 300hp. Compared to the Epsilon, which had a tandem configuration with the instructor in the back and the student in the front, the Grob 120A-F offers side-by-side seating for better instructor supervision in the early stages of the pilot-training syllabus: this configuration is an obvious advantage, allowing the instructors to easily demonstrate manoeuvres while keeping a watchful eye on the actions of their students. Detractors argue that the Epsilon provided a more fighter-like seating arrangement, especially when training navigators. While the Epsilon was made of aluminium and of various alloys, the Grob 120A-F is a much more modern design that extensively relies on composite materials that are both light and strong.

In AAE service, the Grob 120A-F has proved to be a reliable and efficient basic trainer, although a crash was recorded in December 2022. Three squadrons of the École de l'Aviation de chasse 00.315, Escadrons d'Instruction en Vol 2/12 'Picardie' and 4/11 'Jura' and Standardisation and Evaluation Squadron 1/11 'Roussillon', operate the type as part of a shared fleet.

At the time of writing a competition had been launched to replace both the Grob 120A-F and the Cirrus SR20/22 with a new, more advanced trainer. Operating a common type would facilitate maintenance and support and ease training. Contenders include the Grob 120TP, the PC-7 Mk II and the Diamond Dart 450. Compared with both earlier designs, all these turbine-powered aircraft would offer expanded capabilities and performance levels.

The Armée de l'Air was among the first air forces to select the Grob 120 as its next elementary trainer.

In this photo taken at Cognac, in 2018, Grob 120s share the ramp with Epsilons.

This Grob 120 next to a Grob 120TP demonstrator are photographed at Orange Air Base.

This Grob 120 has just landed at Luxeuil, in September 2021. The side-by-side configuration is an advantage for initial training.

Extra 330SC/LC

Stationed in Base Aérienne 701 Salon-de-Provence, the prestigious Équipe de Voltige de l'Armée de l'Air (EVAA) is a unique unit among European military air arms, specialising in sporting events and airshows in France and further afield. It takes part in high-end national and international aerobatics competitions: at the time of writing, five of its pilots had become individual World Aerobatics Champions in 1990, 2009, 2013, 2015 and 2022. EVAA pilots have also become team world champions as part of various French national aerobatics teams (each composed of three civilian and military pilots) in 1990, 2007, 2009, 2013, 2015, 2017, 2019 and 2022. Over the years, EVAA pilots have also won numerous medals during the successive European aerobatics championships.

The EVAA regularly participates in airshows to promote the image of the AAE to facilitate recruiting. Thanks to its light aircraft, EVAA aircrews can display their skills and represent the AAE at the smaller venues where fast jets cannot land, thus allowing the public to interact with AAE personnel, even at the smallest events organised by local air clubs, often on grass runways. The EVAA has recently introduced a two-ship display that includes advanced aerobatics manoeuvres, more attractive to the general public than singletons.

The accident that tragically caused the death of Capitaine Jean-Michel Delorme, in August 2005, led to the grounding and ultimately, the premature retirement of the Mudry Cap 232 aerobatics competition aircraft then in use with the EVAA. The Armée de l'Air soon initiated the search for a replacement

This single-seat Extra 330 lifts off from the runway at Luxeuil for a rehearsal prior to an airshow, in September 2021.

and the German-made Extra 330SC/LC was eventually selected, entering service with the unit in 2008. The Extra 330 was delivered in two variants, single-seat 330SC (two) and two-seat 330LC (one), both powered by a 330-hp Lycoming flat-six piston engine. The Extra 330 is a sprightly performer optimised for international competitions, offering eye-watering agility and excellent performance, even in the vertical thanks to an advantageous power to weight ratio and a very light carbon-fibre airframe: compared to the Cap 232, it is 10 per cent more powerful and 10 per cent lighter. It is cleared to +/-10 g (+/- 8 g for the two-seater with two persons on board) and it can roll at up to 420 degrees per second thanks to extremely large ailerons. Taking advantage of the gyroscopic forces provided by its propeller, the Extra 330 can perform extreme manoeuvres such as the avalanche or the inverted snap roll and is powerful enough to stop in the air and hover like a helicopter, with the airframe literally hung beneath its propeller. For airshows, the EVAA Extra 330s are equipped with smoke generators that help better materialise their impressive trajectories and manoeuvres, allowing the members of the public to better see the path followed by the pilot and better understand how tight the unlimited aerobatics turns, bunts or Humpty bumps really are.

With their new mount, the EVAA has won the most highly prized medals in the field of aerobatics. Although only three aircraft were purchased, they are watched each year by a huge number of spectators and have become incredibly popular. The sole two-seat Extra 330LC was unfortunately heavily damaged and subsequently written off after a forced landing in October 2021. The AAE is now mulling the procurement of a replacement aircraft.

Without doubt, the Extra 330 is the most agile aircraft in service with the French Air and Space Force.

The Extra 330 is both powerful and light, allowing manoeuvres to be easily flown in the vertical.

This Extra 330SC overflies a glacier in the Alps, in November 2009.

The Extra 330s are equipped with a smoke generator to help the public appreciate and memorise their manoeuvres at airshows.

Miscellaneous light training aircraft

The AAE operates a whole range of other light aircraft, motor gliders and gliders in a wide variety of roles. For example, the Cirrus SR20/22 four-seat tourer has been utilised for elementary training at Base Aérienne 701 Salon-de-Provence since 2012. The SR20 provides future pilots with initial training and grading while the more powerful and more advanced SR22 is used for the elementary training of future navigators / weapon system operators. A total of 17 privately-owned military-operated SR20/SR22s are in service. They have replaced the Socata TB10 and TB20 light trainers that had previously been used in that role. One has been lost after engine failure but the crew was saved by the Cirrus fully-integrated airframe parachute system, a whole-plane parachute recovery system designed to safely bring the entire aircraft down (but not necessarily for the prevention of damage to the airframe). At the time of writing a competition had been launched to replace both the Cirrus SR20/22 and the Grob 120A-F with a new trainer.

The nimble Jodel D140 is one of the oldest types in AAE service. Designed in the '50s and powered by an 180-hp Lycoming engine, the D140 has proved to be a very popular light aircraft of wooden construction that has seen extensive service with aeroclubs and private owners in France and abroad. Its cabin can accommodate up to four people. The Armée de l'Air has purchased two variants of the Jodel: 18 D140E Mousquetaires for aircrew training and liaison missions and 15 D140R Abeilles, mainly for glider towing at the various gliding schools. Compared to the Mousquetaire, the Abeille has a narrower rear fuselage and a panoramic canopy that give the pilot better visibility to the rear, an obvious advantage when towing a glider. In all, 17 D140s are still fielded by the AAE today and will remain in service for the foreseeable future. The remaining D140Es have been upgraded to D140ER for gliding towing.

Five Super Dimona HK36 motor-gliders are in service for training missions within a number of AAE gliding schools. These side-by-side aircraft produced by Diamond Aircraft, in Austria, are used from Romorantin, Ambérieu, Saintes and Salon-de-Provence for gliding training and glider towing. They are powered by a 115hp Rotax engine that can be switched off in flight and their airframes are made of composite materials that are both light and strong to guarantee the highest level of performance. Various other types of unpowered gliders are also in AAE service, including the two-seat Duo Discus for competitions, the two-seat Marianne C201 for glider training and the single-seat Pégase for long-range soaring.

A small number of leased light aircraft are operated by the SARAAs (Sections Aériennes de Réserve de l'Armée de l'Air, reserve air force air sections) located at nearly AAE air bases. Usually owned by local air clubs, they are chartered by the AAE as and when required. They undertake a variety of training tasks. For example, they regularly simulate slow-movers that enter prohibited areas, forcing the higher echelons to launch 'T-Scrambles' (training scrambles), with Mirage 2000-5F or Rafale QRA (Quick Reaction Alert) fighters and/or Fennec helicopters to intercept and identify them.

Although not part of the AAE inventory, it is worth mentioning in this chapter the Eurocopter EC120NHE Calliope used by the French Army Aviation School in Dax, in south-west France, for basic helicopter flying training. The school is, in fact, a joint establishment that trains all future military and gendarmerie rotorcraft pilots, plus a number of foreign pilots each year. The EC120NHE (Nouvel Hélicoptère École, or new trainer helicopter), a military variant of the EC120 Colibri, has replaced the Gazelle in that role in the early 2010s, helping train pilots on a modern helicopter type fitted with the latest generation of systems, including an advanced avionics suite and a glass cockpit more representative of the modern helicopters that are being introduced in growing numbers in the French Armed Forces. Future rotary pilots also undertake mountain flying training on the Calliope out of the Centre de Vol Montagne (mountain flying centre), an army establishment at Saillagouse, in the Pyrenees.

The Cirrus has proved to be well adapted for the basic trainer role.

A Cirrus comes in to land at Luxeuil air base in September 2021.

The selection of a civilian touring aircraft as a basic trainer has helped keep pilot-training costs under tight control.

The Jodel D140 has proved to be one of the most successful aircraft in service with the French Air Force.

This Jodel D140E is a Mousquetaire training and liaison variant. It has been modified for glider towing.

The civilian EC120NHE is used by the French Army at Dax to train AAE rotary wing pilots.

This Piper Archer operated by a SARAA is photographed on the apron at Solenzara.

Chapter 6
Surface-to-Air Missile Systems

Thales Crotale NG air-defence systems

As a result of the end of the Cold War and of the ensuing budget cuts as part of the so-called 'peace dividends', a sweeping reorganisation of all the Armée de l'Air and Armée de Terre air-defence units was initiated in France in the 1990s. The outcome was a re-allocation of assets and a sharp reduction in firepower: the Army relinquished its longer-range surface-to-air missiles' (SAM) systems while the Armée de l'Air gave up all the shorter-range air-defence missiles and guns to focus on the medium and long-range segment. As a result, the Army's Hawks and Roland surface-to-air missiles were prematurely sent to the scrap yard, leaving only Mistral short-range missile systems and 20mm cannons to serve with the French Army's surviving air-defence artillery batteries. Armée de l'Air Aspic short-range air-defence launchers and 20mm guns were all withdrawn from use and its Mistrals were transferred on to the Army, leaving only the Crotale NG and the Mamba in service with the AAE. With these lethal systems, the French Air and Space Force is well equipped with layered systems to protect its air bases and other critical installations – command posts, ammunition dumps, harbours, etc. – against various types of threats, including cruise missiles, fighters, helicopters, unmanned aerial vehicles and short-range ballistic missiles (those that can reach targets 300–600km / 186–373 miles away). The Crotales and Mambas are connected via datalink to boost their effectiveness.

All French Air Force Mistral missile launchers that defended air bases were transferred to the French Army.

The Crotale NG is mainly used as a gap-filler and for the defence of high-value bases.

The 20mm cannons that were in widespread use during the Cold War were withdrawn from service in the 2000s/2010s.

The Aspic operator used a helmet-mounted display for weapon aiming.

The Aspic launcher was developed to engage threats at short ranges and supplement the Crotale systems.

The Crotale NG is the final evolution of a long series of Crotale variants. For decades, the Armée de l'Air has operated the Crotale family of short-range SAM systems that had been initially developed in the late '60s by Thomson-CSF for South Africa (locally known as Cactus). At the peak of the East/West confrontation, the Crotale was purchased in large quantities by the Armée de l'Air to provide short-range protection against low-flying jets such as the Su-22 Fitter, the MiG-27 Flogger and the Su-24 Fencer that had become the standard strike fighters among Warsaw Pact countries. During the Cold War, nearly all French Air Force bases had a dedicated squadron equipped with Crotales and 20mm point air-defence cannons.

Today, all 12 Crotale NG (New Generation) towed, air-deployable launchers / trailers remain in AAE service. This variant was specifically developed in the '90s to engage at longer distances the most recent threats, including pop-up helicopters and cruise missiles that were introduced in growing numbers by an expanding number of nations. Each of these trailers, now towed by a Renault Kerax truck, is fitted with a 360-degree surveillance radar, a tracking radar, an electro-optics/infrared day/night identification and tracking suite and with up to eight VT1 missile-launching tubes. Their VT1 trisonic missiles are able to destroy targets at distances of up to 13km (8 miles). The Crotale NG systems that can engage targets from very low level up to 6,000m (19,685 feet) are now mainly used for point air defence and as gap fillers to supplement the much more capable Mamba. Alongside the Mamba, they provide a layered defence against a large range of airborne threats, the Crotale NG being capable of destroying small UAVs and extremely agile and very fast targets. All air-defence assets are coordinated by dedicated command posts relying on datalinks to allocate targets to the various interceptors and SAM systems. At the time of writing, it had been announced that two Crotale NG systems had been delivered to Ukraine to boost the surface-to-air capabilities of the Ukrainian armed forces.

The Aspic launchers were withdrawn from use prematurely and the vehicles transformed into force protection patrol vehicles.

Crotale NG systems are installed in trailers that can easily be moved around the battlefield.

Eurosam SAMP/T Mamba

The SAMP/T (Sol-Air Moyenne Portée/Terrestre, or land-based medium-range surface-to-air) Mamba has become the AAE's main air-defence system. This advanced system was jointly developed by France and Italy as part of the Eurosam consortium. It utilises the same Aster 30 missiles also in service with the French and Italian navies, with the Royal Navy on board Type 45 destroyers and with various export customers. The requirement was for a very-long range system that would secure a huge area thanks to the latest generation of sensors and of powerful fire-and-forget missiles that could be fired in salvoes to defeat multi-axis saturating attacks. The SAMP/T was intended to provide massively expanded capabilities compared to the Crotale, the Crotale NG, the Roland and the Hawk.

The Mamba entered service in France in the late 2000s and eight systems are now fully operational with five Escadrons de Défense Sol-Air at Avord, Istres, Luxeuil, Mont-de-Marsan and Saint-Dizier, and with the Centre de Formation à la Défense Sol-Air, at Avord. At the time of writing, the transfer of one of these systems to Ukraine was under discussion.

Each Mamba firing section is composed of an Arabel panoramic electronic / mechanical scanning radar module for long-range detection and tracking, of a command post, of a generator module and of four launchers, each with up to eight Aster 30 missiles. These vertically-launched missiles can defeat air-breathing targets and short-range ballistic missiles at distances exceeding 100km (62 miles) and at altitudes of more than 20km (65,000 feet). Their no-escape range against a hard-manoeuvring fighter is said to be around 70km (43 miles), although all precise data is classified. This means that each Mamba system provides true area defence capabilities over 360 degrees. The launchers can be positioned away from the command post to defend an even wider area. The various modules and launchers are all mounted on 8 x 8 Renault Kerax all-terrain trucks for tactical mobility. The entire system can be air-transported by A400M Atlas airlifters.

All the AAE Mamba systems will be soon upgraded with the Aster 30 Block 1 NT (New Technologies) missile that will offer massively expanded reach against more lethal ballistic missiles (typically those

A SAMP/T missile launcher and a Rafale B omnirole fighter share the ramp at Saint-Dizier. They both represent the most advanced air-defence assets in France.

with a range of 1,300–1,600 km / 808–1,000 miles). A new Ka-band seeker and a new central computer will be among a series of modifications that will provide improved interception capabilities to the Aster 30 Block 1 NT. This new missile will be used in conjunction with the new Thales Ground Fire 300 radar, which will replace the current Arabel and provide considerably improved detection and tracking capabilities. While the Arabel has a range of more than 100km (62 miles), the Ground Fire 300 will spot and track targets at distances exceeding 300km (186 miles), a massive boost in operational capabilities. With its ability to reach altitudes of more than 70km (230,000 feet), the future Aster Block 2 missile now envisioned for the next round of upgrade of the Mamba system will be capable of engaging the latest threats, including hypersonic missiles that cruise at more than Mach 7.

In addition to its SAM systems, the AAE is now investing in various technical solutions to defeat the drones and micro-drones used in ever increasing numbers by potentially hostile forces. Various types of systems are already fielded or under development. They rely on a range of innovative solutions for hard and soft kills, including electronic warfare and other micro-drones used as anti-drones interceptors.

The Arabel panoramic electronic / mechanical scanning radar module used for long-range detection and tracking of aerial targets.

All modules of the SAMP/T Mamba systems are mounted on Kerax 8 x 8 trucks, providing good all-terrain mobility.

A programme to replace the Mamba's current Arabel radar with the longer-range Ground Fire 300 radar has been launched.

Left: A Mamba command post photographed at Mont-de-Marsan air base.

Below: Mamba-firing sections are designed to protect high value targets such as air bases, harbours or logistic bases.

A view of a firing control workstation inside a Mamba command post. Mamba systems are both rugged and high-tech.

A Mamba command post and a missile launcher at their Saint-Dizier home base. In real life, they would be positioned hundreds of metres away from each other.

Glossary

Armée de l'Air et de l'Espace: French Air and Space Force
Forces Aériennes Stratégiques: strategic air command
Brigade des Forces Spéciales Air: AAE's special forces brigade
Centre d'Expertise Aérienne Militaire: military air expertise centre
Centre de Formation à la Défense Sol-Air: surface-to-air missile training centre
Centre d'Entraînement des Équipages de Transport: transport crew training centre
Centre d'Instruction des Équipages d'Hélicoptères: helicopter crew training centre
Direction des Centres d'Expérimentations Nucléaires: nuclear experiment centres directorate
École de l'Aviation de chasse: fighter training school
École de l'Aviation de Transport: transport aviation school
Escadron de Chasse: fighter squadron
Escadron de Chasse et d'Expérimentation: operational evaluation and fighter squadron
Escadron de Transformation: conversion training squadron
Escadron Électronique Aéroporté: airborne electronic squadron
Escadron de Détection et de Contrôle Aéroportés: airborne detection and control squadron
Escadron de Drones: UAV squadron
Escadron de Transition Opérationnelle Drone: UAV operational conversion squadron
Escadron de Ravitaillement en Vol: air-refuelling squadron
Escadron de Ravitaillement en Vol et de Transport Stratégique: in-flight refuelling and strategic transport squadron
Escadron de Transport: transport squadron
Escadron de Transport, d'Entraînement et de Calibration: transport, training and calibration squadron
Escadron de Transport d'Outre-Mer: overseas transport squadron
Escadron d'Hélicoptères: helicopter squadron
Escadron d'Entraînement: training squadron
Escadron d'Instruction en Vol: flight instruction squadron
Escadron de Défense Sol-Air: surface-to-air missile squadron